52
Topics in Heterocyclic Chemistry

Aims and Scope

The series Topics in Heterocyclic Chemistry presents critical reviews on present and future trends in the research of heterocyclic compounds. Overall the scope is to cover topics dealing with all areas within heterocyclic chemistry, both experimental and theoretical, of interest to the general heterocyclic chemistry community.

The series consists of topic related volumes edited by renowned editors with contributions of experts in the field. All chapters from Topics in Heterocyclic Chemistry are published OnlineFirst with an individual DOI. In references, Topics in Heterocyclic Chemistry is abbreviated as Top Heterocycl Chem and cited as a journal.

More information about this series at http://www.springer.com/series/7081

Viktor Krchňák
Editor

Solid-Phase Synthesis of Nitrogenous Heterocycles

With contributions by

C.M.L. Delpiccolo · F. Diness · V. Fülöpová · V. Krchňák ·
A. La-Venia · E.G. Mata · M. Meldal · E. Schütznerová ·
G.A. Slough · M. Soural · Y. Wang

Springer

Editor
Viktor Krchňák
University of Notre Dame
Notre Dame, Indiana
USA

Palacky University
Olomouc, Czech Republic

ISSN 1861-9282 ISSN 1861-9290 (electronic)
Topics in Heterocyclic Chemistry
ISBN 978-3-319-58615-1 ISBN 978-3-319-58616-8 (eBook)
DOI 10.1007/978-3-319-58616-8

Library of Congress Control Number: 2017942797

Printed on acid-free paper

This Springer imprint is published by Springer Nature
The registered company is Springer International Publishing AG
The registered company address is: Gewerbestrasse 11, 6330 Cham, Switzerland

Preface

In 1963 Robert Bruce Merrifield forever changed peptide synthesis when he revealed the concept of solid-phase (SP) peptide synthesis [1]. Merrifield's approach simplified isolation of intermediates during traditional stepwise peptide synthesis. The carboxy-terminal *N*-protected amino acid was covalently attached to insoluble polymeric support. Resin beads of polystyrene cross-linked with divinylbenzene served as the solid phase. Even to this day, this support is the most frequently used in all aspects of SP synthesis. In practice, SP peptide synthesis consisted of repetitive use of two steps: removing the amino acid *N*-protecting group and coupling of the next *N*-protected amino acid. After each step, simply washing the resin beads can isolate the resin-bound intermediates.

The potential of SP synthesis for general organic chemistry was soon recognized, including Henry Rapoport [2] who realized the unique advantages of SP synthesis. However, it was not until the era of combinatorial chemistry, started in the early 1990s by Kit Lam [3] and Richard A. Houghten [4], that SP organic synthesis started to flourish. Pooling of resin-bound intermediates enabled synthesis of large chemical libraries. The first SP heterocycle synthesis was reported from Jonathan A. Elman's laboratory [5].

SP synthesis of heterocycles expanded Merrifield's original idea by applying a sequence of unrelated organic reactions on resin-bound intermediates. Obviously, each step of the SP heterocycle synthesis needs to be fine-tuned to provide the highest possible purity of resin-bound intermediates, because the intermediates cannot be purified. The successful SP synthesis needs to master not only two reactions used for peptide synthesis (acylation and *N*-protecting group cleavage), but to optimize all reactions involved in synthesis.

Critical advantages of SP synthesis include very efficient isolation of intermediates by simply washing reagents away from the resin beads with organic solvents. Unlike synthesis in solution, the isolation timing is therefore very predictable and allows multiple syntheses to be conducted at the same time. Because of simple reaction solvent removal, high boiling solvents such as DMF and DMSO can beneficially be used without the need for evaporation.

On the other hand, SP synthesis requires detailed optimization of reaction conditions for each step of the synthesis, excess of reagents to drive the reaction to completion, and excludes chemical transformations using heterogeneous components (e.g. catalyst).

The five chapters in this book describe different facets of SP heterocycle synthesis. The first chapter by Greg A. Slough is included to inspire novices in SP synthesis and describes details of each of the SP synthesis aspect. He has proven that SP synthesis can be carried out in any organic chemistry laboratory, without significant investment, using a plastic syringe equipped with a porous disc served as a reaction vessel for manual SP synthesis.

Veronika Fülöpová and Miroslav Soural summarize the rich literature dedicated to the synthesis of seven-membered nitrogenous heterocycles including notoriously known pharmacophores such as benzodiazepines. This chapter shows that numerous different routes are applicable for SP synthesis of the same type of heterocycle, and chemists can select a route best suited for a given project's needs.

Morten Meldal's laboratory dedicates substantial effort to developing very powerful cyclic iminium – nucleophilic addition cascade reactions. Together with Frederik Diness and Yuanyuan Wang as co-authors they contribute a chapter documenting the power of this cascade reaction for SP synthesis of diverse fused and bridged molecular scaffolds applicable as peptidomimetics.

Agustina La-Venia, Carina M. L. Delpiccolo and Ernesto G. Mata focus on methodology and their chapter discusses metal-mediated synthesis of heterocyclic compounds. This area, in particular, is a growth area in heterocyclic chemistry, and we can expect more critical contributions in coming years.

The last contribution by Eva Schütznerová and Viktor Krchňák describes syntheses of diverse nitrogenous heterocycles based on C-arylation reaction of 2-nitrobenzene sulfonamides. Interestingly, these synthetic routes were developed for SP in an arena with no precedents for synthesis in solution.

To conclude, individual chapters in this book document that SP synthesis can provide the ultimate solution to very efficient synthesis of structurally diverse heterocyclic compounds.

Notre Dame, IN, USA Viktor Krchňák

References

1. Merrifield RB (1963) Solid phase peptide synthesis. I. The synthesis of a tetrapeptide. J Am Chem Soc 85:2149–2154
2. Crowley JI, Rapoport H (1976) Solid-phase organic synthesis: novelty of fundamental concept? Acc Chem Res 9:135–144
3. Lam KS, Salmon SE, Hersh EM, Hruby VJ, Kazmierski WM, Knapp RJ (1991) A new type of synthetic peptide library for identifying ligand-binding activity. Nature 354:82–84

4. Houghten RA, Pinilla C, Blondelle SE, Appel JR, Dooley CT, Cuervo JH (1991) Generation and use of synthetic peptide combinatorial libraries for basic research and drug discovery. Nature 354:84–86
5. Bunin BA, Ellman JA (1992) A general and expedient method for the solid phase synthesis of 1,4-benzodiazepine derivatives. J Am Chem Soc 114:10997–10998

Contents

Top Heterocycl Chem (2017) 52: 1–22
DOI: 10.1007/7081_2016_3

Published online: 13 April 2017

Solid-Phase Synthesis of Heterocycles in Practice

Greg A. Slough

Abstract To demonstrate the simplicity of adopting solid-phase synthesis of heterocycles by novices, we developed an appropriate education module (3 weeks in length) for the second-year undergraduate Organic Chemistry laboratory that effectively introduced solid-phase synthesis into the undergraduate curriculum. Starting with Wang resin, a four-step synthesis involving a 2-nitrobenzenesulfonamide intermediate and two points of chemical diversity produced a ten compound library of indazole 1-oxides in good yield. Cleavage of the final resin construct with 50% TFA/DCM provided each student with a sample of their unique compound. Four short instructional video clips supplement the laboratory activities described in this chapter.

Keywords Crosslinked polystyrene • Hands-on learning • Indazole 1-oxide • Instructional module • Linker • Second-year undergraduate • Solid-phase stoichiometry • Solid-phase synthesis

Contents

Supplementary material is available in the online version of this chapter at 10.1007/7081_2016_3. Videos can also be accessed at http://www.springerimages.com/videos/.

G.A. Slough (✉)
Chemistry Department, Kalamazoo College, Kalamazoo, MI 49006, USA
e-mail: greg.slough@kzoo.edu

1 Background

1.1 Case Study

A 3-week laboratory instructional module used at Kalamazoo College, Kalamazoo, Michigan, trains undergraduates in the art of solid-phase synthesis. This module utilizes reading assignments, instructional videos, and discovery-based experiments to teach students important aspects of combinatorial solid-phase synthesis. The overall goal of this module is to prepare a small diversomer library (two points of diversity) of indazole 1-oxides. While this chapter will not discuss organic heterocycles and their role in medicinal science, pre-laboratory talks address aspects of these important chemicals and provide an introduction to heterocycle nomenclature. Learning outcomes for this laboratory module attempt to connect the chemistry of heterocycles to chemical principles studied in carbocyclic chemistry and general functional group reactivity. Lastly, this discovery-based exercise provides a means to modernize the chemistry curriculum and incorporate maturing synthetic methodologies that will be relevant to a student's professional career.

Non-practitioners of combinatorial or solid-phase synthesis often have the misconception that this methodology is expensive and/or difficult to master. This case study suggests that neither criticism is valid. The versatility and speed of solid-phase chemistry makes it ideally situated over a range of applications from an instructional platform to advanced research laboratories. Polypropylene syringes with porous discs serve as simple, inexpensive, and disposable reaction vessels [1]. The synthesis can be carried out manually, simply by drawing the solution/solvent into a syringe with resin, shaking the resin slurry, and dispensing the solution/solvent. For parallel synthesis in several reaction vessels (syringes), it is advantageous to integrate common synthetic steps such as washing resin beads. The Torviq Domino-Block method [2] (Fig. 1), used in this case study, requires an initial investment of ~$3,000 for a 24 person laboratory section. After this initial investment, year-to-year expenses are only slightly more expensive than traditional laboratory exercises. Most of this added expense (~$1 per student) comes from the purchase of commercial solid-phase resin. While the cost of commercial Wang-imidate resin is appreciable, there are both cost and reactivity advantages to prepare this same resin in one step from less expensive Wang resin. Common equipment, such as titer plate shakers, shaker-rotisseries, and gas cylinder regulars, are in the inventory of many science departments which decreases the startup cost. During the developmental phase of this module, Kalamazoo College trained 55–65 organic

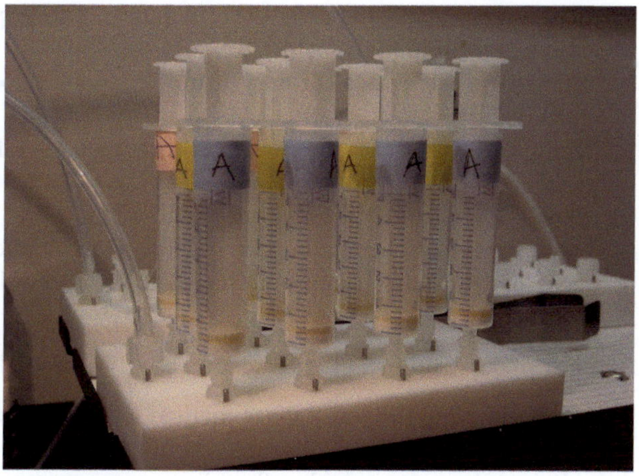

Fig. 1 Torviq Domino-Block reaction vessel

chemistry students in three laboratory sections annually, and after the first year, the laboratory instructional budgets nearly returned to previous levels.

With respect to mastery, Organic Chemistry students conducted 16 solution-phase experiments or reactions and had basic training in isolation methods, reflux-distillation techniques, and spectroscopy before starting the solid-phase synthesis module. Class discussions integrated concepts such as retrosynthesis and linear/convergent syntheses, so stepwise laboratory preparations seemed natural to most students.

Prior to this instructional module, a take-home survey of students (3 years of sampling) demonstrated only cursory or "Wikipedia-level" knowledge of solid-phase synthesis. Illustrative responses to the question "Provide your best definition of 'solid-phase synthesis'" included:

- "In biology, I know it is used to synthesize peptides in a specific alignment."
- "Solid-phase synthesis is when organic molecules are bound to an insoluble support so any unreacted reagents can be removed easily."
- "Bonding molecules together on a bead. The process stops after each step."
- "Molecules are temporarily bound to an insoluble support. Easily purified by filtration."

While these responses contain facets associated with solid-phase reactions, not one student identified equilibria or kinetics/thermodynamic features that are greatly enhanced in solid-phase reactions. In addition, no one identified cleavage of the final organic product from the polymeric scaffold as a logical consequence of this technology.

Post-module surveys assessed outcomes for students in terms of advantage(s) and/or disadvantage(s) of solid-phase synthesis, and in terms of new understandings of chemical reactivity. Responses to the advantage/disadvantage prompt included:

- "The greatest advantage is the efficiency. The experiment is fast-paced and products form in high-yield."
- "A gigantic chemical library can be synthesized easily (automated) and quickly."
- "Simple filtration as the purification methods – allows for the use of excess reagents."
- "Not able to use NMR as an analytical method."
- "Solid-phase synthesis can only prepare a small amount of product."
- "It is hard to tell if a mistake has been made until analysis at the end of the scheme."

New understandings of chemical reactivity included:

- "By employing the solid-phase method we learned about driving reactions with excess chemical and exploring reactions that are not readily observed in the solution phase (even with reflux)."
- "Kinetic and thermodynamic conditions can form different products."
- "It is important to consider acidity of compounds when determining how an rxn [sic] will occur – whether an ester or ketone is more acidic."

The sophistication of student responses on post-module survey questions represents learning outcomes that are more dynamic and more effective than any other synthesis module used previously. While it is hard to pinpoint exact features of this module that facilitate learning, the ability to optimize reactions, adjust reagent concentrations, adjust substrate structure in rapid succession fit with a model of maturing scientific thinking.

Looking at summative questions, such as: "After completing the series of experiments related to solid-phase synthesis project, I feel that I learned an important technique in organic chemistry," showed that 98% (3-year average, $n = 172$) strongly agreed or agreed with this comment. As to whether they enjoyed doing solid-phase synthesis more than other types of experiments done during the laboratory sequence, the responses were more mixed.

1.2 Context of the Chapter

Experimental and experiential data from this solid-phase training module demonstrate that skills and techniques once thought to be esoteric and specialized to the solid-phase artesian are tractable teaching and learning goals for Organic Chemistry students. Combinatorial syntheses on solid-phase supports can be transplanted to other undergraduate chemistry curricula. A few other institutions [3] [See reference herein for the development of the Distributed Drug Discovery (D3) model, 4] and other academic disciplines [5] have also begun to incorporate combinatorial solid-phase synthesis into undergraduate programs. These recent contributions to the undergraduate laboratory curriculum contextualize this training in terms of an introduction to chemical research. Several professional editorials [6, 7] and governmental advisory papers [8, 9] document positive impact of scientific research experiences during the undergraduate studies.

With careful design, many areas of technology and engineering, apart from heterocycle synthesis, could be developed into an effective training pedagogy. One common issue unrelated to solid-phase technology is workload for laboratory stockrooms and technical support staff. Our experience at Kalamazoo College has shown this concern to be unfounded. Once the mechanical shakers, Domino Block synthesizers, and nitrogen gas cylinders/regulators/tubing (two each of this equipment for a 24 person lab) have been positioned in the laboratory space, the work of stockroom staff actually decreases. Every week, two or three new solutions of reagent must be prepared and stored in septum-sealed bottles. Because almost anchor sitess in the synthetic scheme are run in excess of a reagent, the exact concentration is not particularly important. However, it is important to keep the concentrations above ~0.2 M. Below 0.1 M the reactions are sluggish and often do not complete. This lower limit is particularly noticeable with resins with low reactive site loading (<0.1 mmol/g).

The remainder of this chapter will be written at a developmental level suitable for second-year undergraduate chemistry students, and much of this material could be amenable for a laboratory manual. Unlike peer-reviewed journals on chemical education where procedures, training methodologies, and student outcomes are the centerpieces, this chapter identifies and annotates three problematic areas that often puzzle students: (1) the influence of polymer swelling (solvent absorption) in solid-phase synthesis, (2) the role of chemical linkers, and (3) the understanding of stoichiometry in the presences of unreactive polymer. Notes and discussions of these problem areas are provided as both learning and teaching resources.

The organization of this chapter will start with general principles of solid-phase synthesis with reference to Merrifield's pioneering work and then progress to stoichiometry problems based on resin loading levels. A full synthetic procedure of indazole 1-oxides with appropriate timelines is then provided as a guide for instructing a second-year undergraduate organic chemistry laboratory.

2 General Principles of Solid-Phase Synthesis

2.1 Introduction

The development of the "perfect" chemical that possesses "perfect" chemical and physical properties is a recurring theme in modern science and technology. Whether it be pharmaceutical companies searching for structure activity relationships that meet demanding requirements of health safety and efficacy or electronics companies looking for the next "best" organic LED (OLED), the thirst for new advanced chemicals and materials is insatiable. As demand grows for new compounds, manual (human), semi-automated, and automated synthetic approaches must be learned, studied, and evaluated to gain competence in the field of Organic Chemistry.

This new professional environment is substantially different than what R. B. Merrifield found in the field of enzymology in the 1950–1960s. Demand for small peptides of known amino acid constitution dominated almost all aspects of this discipline. Enzymology and biochemistry needed a flexible supply chain of peptides in order to develop. To be clear, recombinant protein expression as a technique was not even dreamed about when Merrifield surveyed the enzymology/biochemistry landscape. Merrifield observed one other thing; linear-solution-phase syntheses of peptides were often cumbersome, low yield endeavors. In Merrifield's first paper describing his new synthetic method, ultimately called *solid-phase peptide synthesis*, he rationalized his approach by pointing out that: "these [existing] procedures are not ideally suited to the synthesis of long chain polypeptides because the technical difficulties with solubility and purification become formidable as the number of amino acid residues increases." [10] Merrifield's new proposition "depends on the attachment of the first amino acid of the chain to a solid polymer by a covalent bond, the addition of the succeeding amino acids one at a time in a stepwise manner until the desired sequence is assembled, and finally the removal of the peptide from the solid support." [10] Extensive studies with polymeric supports including cellulose, polyvinyl alcohol, polymethacrylate, and sulfonated polystyrene demonstrated that 1–2% divinylbenzene crosslinked polystyrene beads swell into a porous insoluble gel in hydrophobic organic solvents. In addition, the polystyrene beads were rather inert to many organic chemical reagents.

In order to form a point-of-attachment for the first amino acid, Merrifield utilized an electrophilic aromatic substitution reaction to chloromethylate a small fraction of the aromatic rings in the polystyrene backbone (**2**, ~0.5 mmol/grams of resin). Nucleophilic substitution at the chloromethyl group with Cbz-amino acids (Cbz = $PhCH_2OCO$) gave the requisite "anchor bond" that secured the C-terminus amino acid to the resin (**3**, Scheme 1). Merrifield used the term "anchor bond" throughout his peer-reviewed and popular press reports on solid-phase synthesis even though it is not an apt description [11, 12]. Fortunately for Merrifield, high yield reactions from solution-phase chemistry translated smoothly to the solid-phase, and the N-terminus of the first amino acid could be deprotected, and then acylated with the next protected amino acid. Repetition in a predetermined order of amino acids (up to ~30 AA) allowed both natural [13–15] and site-specific peptide mutagens

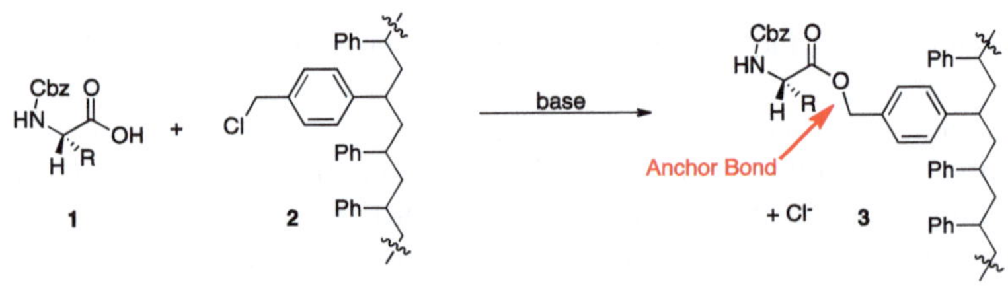

Scheme 1 Merrifield's original substitution reaction with solid-phase resins

[16] to be prepared in sufficient purity and high enough yield to have a dramatic impact on many biological chemistry fields.

From Merrifield's original work, three enduring principles persist into modern solid-phase synthesis:

1. The solid support must be recoverable by simple mechanical/filtration methods at any point of a synthetic scheme or reaction.
2. The solid-phase support must swell in hydrophobic organic solvents allowing bonding-distance approach between the supported organic substrate and a chemical reagent.
3. The supporting polymeric material must be inert to chemical reactions except at specific loci (now known as linkers) where the anchor bond can be formed and broken as desired.

To fully appreciate how these three principles become practical in semi-automated and automated synthesizers, this chapter will briefly explore the nature of polystyrene resins, state-of-the-art linkers that negotiate between a synthetic organic molecule and the solid-phase resin, and reaction stoichiometry in the presence of unreactive polymers.

2.2 *Constitution of the Polystyrene Bead*

The polymerization of styrene monomer has been known for more than a century [17]. In its earliest form this atactic, amorphous polymer possessed many features that are familiar in everyday life. It acts as an insulator, a low electrical conductor, and it has the ability to foam and be molded. With respect to our topic of solid-phase synthesis, simple non-crosslinked polystyrene is completely unsuitable for our considerations because of its solubility in solvents such as acetone, tetrahydrofuran, dichloromethane, and dimethylsulfoxide.

By the mid-1960, when Merrifield started his work, two chemical technologies had been introduced into polystyrene chemistry. First, mixtures of divinylbenzene (1–15% DVB) and styrene co-polymerized efficiently to give crosslinked-polystyrene (PS) beads (**5**, Fig. 2). DVB in this co-polymer adjoins two separate chains of polystyrene and in doing so modestly changes the constitution and physical properties of the solid phase. For example, these co-polymers possess a more rigid morphology, and they are rather insoluble in most solvents. Instead of dissolving, crosslinked PS beads (1–2% DVB) swell due to the absorption and adhesion of solvent molecules within the PS bead. As seen in Fig. 3, swollen PS beads retain their discrete cellular composition, but more importantly, the beads behave more like a sponge than a balloon.

Electrostatic attraction between solvent molecules and the polymeric network determines the size of the solvated bead. The extent of swelling depends on the type of swelling solvent used. For example, tetrahydrofuran and dichloromethane are

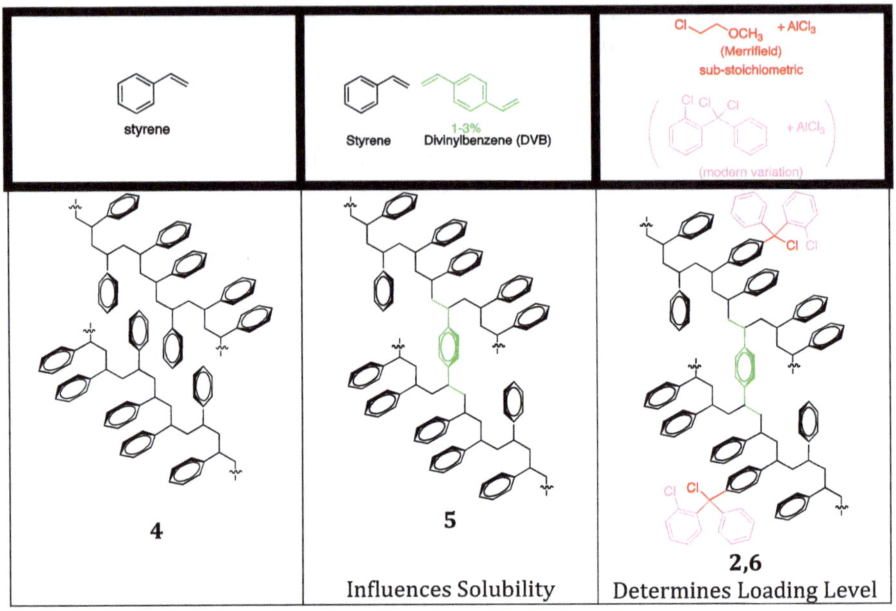

Fig. 2 Polystyrene backbone is the basis for commercial polystyrene and solid-phase synthesis support

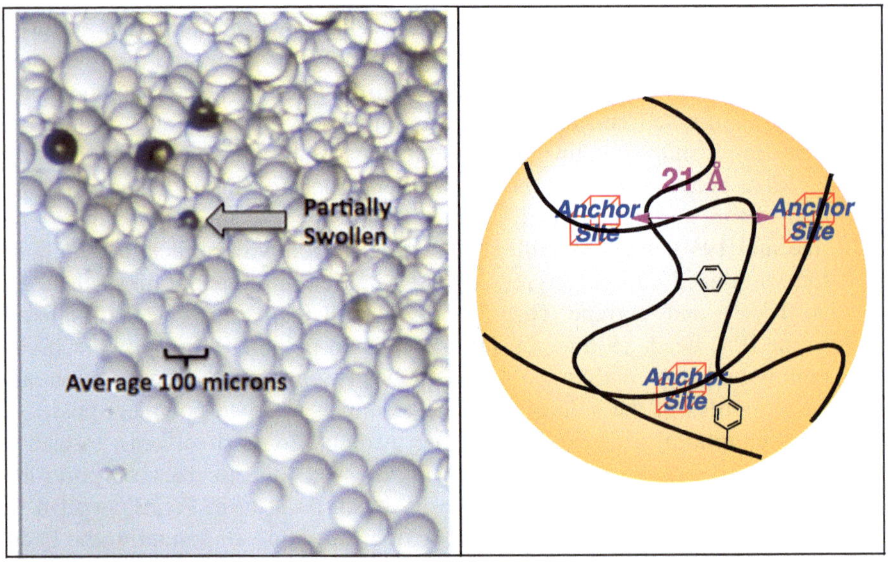

Fig. 3 Swelling characteristic of 1% DVB-styrene polymer beads (**5**) in dichloromethane. Both the micrograph (×25) and illustration pertain to ~1.0 mmol/gram loading level

considered excellent swelling solvents; as such one gram of 1% crosslinked PS resin absorbs about 5.2–5.5 mL of each solvent and on average beads swell to approximately 100 microns in size. Acetonitrile and ethanol are modest swelling solvents absorbing only 3.5–2.0 mL/grams of resin. Water does not swell crosslinked PS resins at all [18].

Our understanding that solvents absorb into pores and crevasses, like a sponge, comes from the fact that anchor sites are primarily located inside the swollen bead. Using simple assumptions based on the volume of a 100-micron bead, the mass of polystyrene in a bead, and the amount of solvent absorbed by the bead, one can estimate that 99% of anchor sites/anchor sites (approximately 50×10^{12} sites) are contained within each polymeric sphere [19, 20]. Applying this assumption, one can estimate that a resin with ~1.0 mmol/g loading has reactive site to reactive site distance of between 20 and 60 Å depending on which geometric model (sphere, cyclinder, cube) is evaluated [At a latter date it was found that higher loading adversely affected coupling at the anchor site, 21]. EPR studies of site–site distance using line broadening of paramagnetic agents (such as TOAC) estimate a distance ~21 Å between reactive sites. This estimate only fits the solvated bead model. A balloon model, with the solvent filling the interior of a polymeric barrier, does not fit these structural features.

Accompanying the invention of crosslinking is the discovery of heterogeneous coordination catalysts that allowed the production of isotactic polystyrene. Tacticity refers to the stereochemical configuration along the polymeric backbone (Fig. 4).

Crosslinking within the polymer, tacticity, and average molecular mass of the polymer sample all have a profound influence on the extent of bead swelling and ultimately on the success of a solid-phase synthesis. For example, if the percent of crosslinking it too high, the insoluble PS beads become fragile and glass-like. These beads crush and pulverize during mechanical filtration. Tacticity also affects swelling characteristics. This is a much more subtle and difficult feature to discern prior to experimentation. Bouillon, Soural, Miller, and Krchnak have documented cases where two commercial resins with identical specifications (bead size, crosslinking, and loading) (Fig. 5), at least to the level of average molecular mass, reported that loading levels, and crosslinking, swell differently in the same solvent. These differences adversely affect reactivity [22].

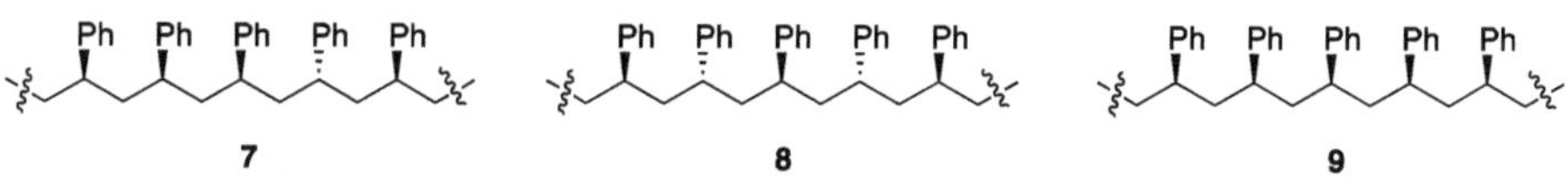

Fig. 4 Tacticity in polymers: atactic – random stereoisomeric centers, **7**, isotactic – alternating stereo-configurations, **8**, syndiotactic – consistent stereo-configurations, **9**

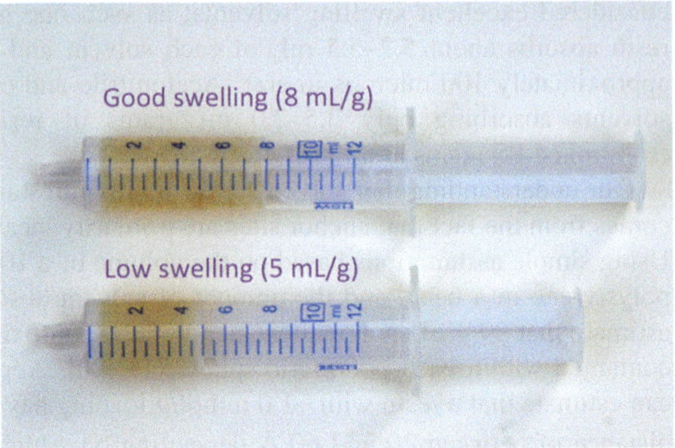

Fig. 5 Crosslinking, polymer tacticity, and average molecular mass influence swell in PS beads. Swelling studies must be completed before processing with a chemical synthesis

2.3 Linkers

The anchor bond is perhaps the most important feature in solid-phase synthesis. It has many fundamental tasks during a successful preparative scheme. This bond must form in very high yield because this yield affects all other chemical yields, and it must be stable enough to endure a variety of chemical conditions. Finally, it must cleave, separating the expected product from the PS resin, under specific conditions, at any point along a synthetic pathway. This is indeed a demanding list of requirements. Merrifield's simplistic approach, as seen in Scheme 1, used the ever-reliable S_N2 reaction to form the anchor bond, but, unfortunately, the anchor bond to the resin proved too unstable during repetitive removal of the Cbz (N-terminal) deprotection, acyl coupling, and neutralizations. Nitration or bromination of the polystyrene support, before the first amino acid anchored, increased acid stability of polymer-supported esters and moderated many reactivity problems. An alternative invention, the Fmoc-amino protecting group, cleavable by a base (piperidine), secured the anchor bond and many more coupling cycles could be achieved. While each of these modifications led to greater and greater peptide complexity, similar alterations proved incompatible with a wider range of organic syntheses.

Viewed from a different perspective, the anchoring benzylic O–C (N–C) bond is too unreactive. As seen in Scheme 2, an immobilized peptide (simplified here to a single amino acid) has two bonds available for hydrolytic cleavage: (1) the O-acyl bond and (2) the benzylic O–C bond. The chemistry of acid derivatives strongly suggests that the O-acyl bond hydrolyzes much faster leaving the original anchor bond intact. If an amino-alcohol (Y = H) is anchored instead, the benzylic O–C bond would be a faster acid mediated cleavage, but overall the reaction would be very slow. The benzylic ester and ether illustrated in these two examples show that simple benzylic O–C bonds are not practical covalent anchors in solid-phase chemistry as they would require very strong acid (e.g., $HF_{(l)}$) to release the product from the resin.

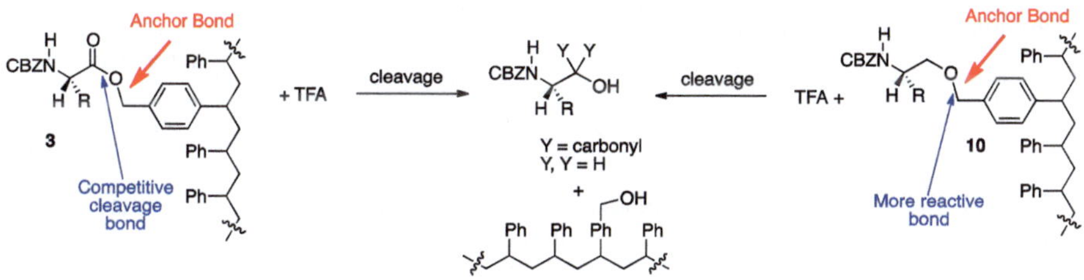

Scheme 2 Functional group at the Anchor Bond influences cleavage site

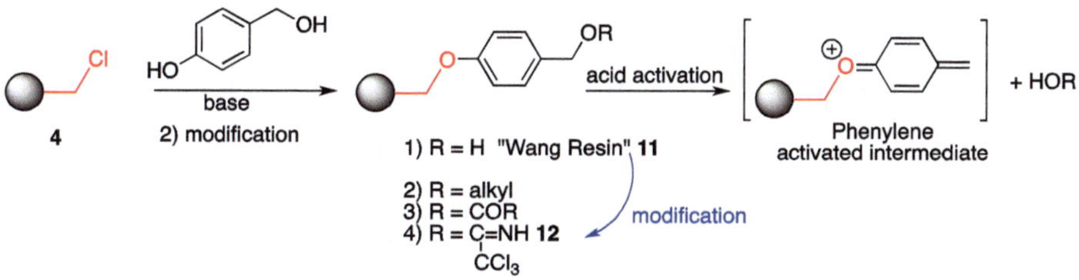

Scheme 3 Substitution of 4-hydroxybenzyl alcohol at the Anchor Bond produces the Wang linker

This conundrum of benzylic esters and ethers forced a rethinking of underlying principles of solid-phase chemistry. One of the first substantive steps in this reappraisal came from the introduction of more selective chemical agents at the original anchor site. Over time this new approach became known as solid-phase linkers. One of the first and most successful strategies came from fundamental studies of Su-Sun Wang [23]. Wang's insight focused on stabilizing positive charge at the benzylic carbon during substitution chemistry. This applies to both the "on" anchor bond and "off" anchor bond cleavage. Wang ultimately settled on 4-hydroxybenzyl alcohol as a capable chemical linker. Substitution of 4-hydroxybenzyl alcohol onto chloromethyl PS-resin yields a solid-phase support, commercialized as Wang resin, which under acid conditions has superior control over the anchor bond. It should be reiterated here that addition of 4-hydroxybenzyl alcohol, in this case, does not add any new loading sites to the resin. Rather it simply modifies existing loading sites.

With the expanded capability of linkers, functional groups such as alcohols, carboxylates, and amines by way of simple carbamates (N–COO) can be added or cleaved under select acid conditions. For example, activation of trichloroimidate Wang resin [24] with BF_3·etherate (Lewis acid) generates a phenylene cation that smoothly couples alcohols, acids, and amides. At the end of a synthesis it is common to cleave the benzylic O–C or N–C with trifluoroacetic acid solution (50% TFA in DCM) (Scheme 3).

Depending on the synthetic application, Wang resin may lack the correct acid activation profile, thus many other chemical linkers have been developed over the past 20–30 years to accommodate a vast array of chemical environments (Fig. 6). The type of functional group being attached determines in which linker to use.

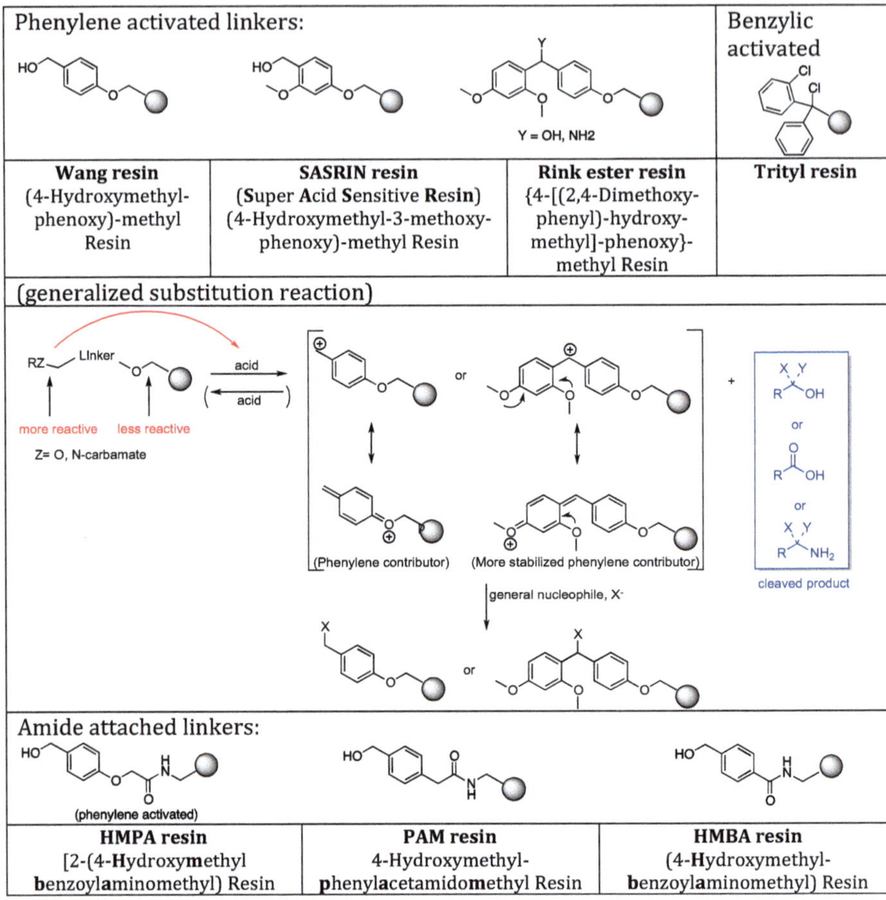

Fig. 6 Examples of commercial linkers on polystyrene resin

As the leaving group ability of a functional group decreases ($RCOO^- > RO^- > RCON^- > RNH^-$) the reactivity of the linker must increase (Wang< SASRIN<Rink). For example, Wang is a reasonable choice for most carboxylates, alcohols, and/or carbamates, while amides and secondary amides normally require either a Rink or BAL linker to facilitate anchor bond formation and anchor bond cleavage. At their very root, however, almost all linkers build on phenylene cation or benzylic cation stabilization for their specific activation.

2.4 Calculating Solid-Phase Stoichiometry

Based on the composition of solid-phase supports, it should be clear that reaction stoichiometry must consider both the mass of unreactive polymeric support and the loading level reported for a commercial resin. For instance, a 1%-DVB-PS resin sold as 1.06 mmol/g of Wang-trichloroimidate should bond 1.06 mmol (theoretical limit) of organic substrate throughout any synthetic scheme. In other words, an advanced synthetic intermediate could be prepared under conventional solution-phase conditions and then substituted onto the Wang-trichloroimidate resin late in the synthesis. This assumes that the synthetic intermediate possesses a competent functional group for substitution. Under theoretically limiting conditions, this new substituted resin should be identical to a synthetic resin prepared entirely on the solid-phase, and the amount of substrate on the resin should be 1.06 mmol assuming limiting yields in each step.

The ability to envision substitution as a route to a synthetic resin provides an unambiguous way to interpret reaction stoichiometry in the presence of unreactive polystyrene. To work through an example, the substitution route in Scheme 4 would take a synthetic intermediate ($C_{16}H_{21}O_2$) and displace trichloroacetamide (CCl_3CONH_2). After this substitution, the only observable physical change would be the mass of the synthetic resin. The myrtenol derivative (245 g/mol) weighs more than trichloroacetamide (162 g/mol), and this difference times 1.06 mmol (resin loading) would increase the mass of the resin by 88 mg. Therefore, in order to measure 1.06 mmol of reaction sites one would need 1.088 grams of resin. Under limiting conditions, the synthetic route in Scheme 4 would produce the same 1.088 grams of resin. Along a synthetic pathway, one must remain cognizant of two things: the mass of the organic fragment in the reagent resin and the mass of the original leaving group on the commercial resin. With these two values, the theoretical mass of the resin can always be determined. For example, the Resin Loading Calculator spreadsheet, Fig. 7, shows that the reagent resin, with 2-phthalimidoethanol attached, requires 1.031 grams of resin to have 1.06 mmol of reactivity. This mass difference, 1.00 g vs 1.031 g, reflects the greater mass of

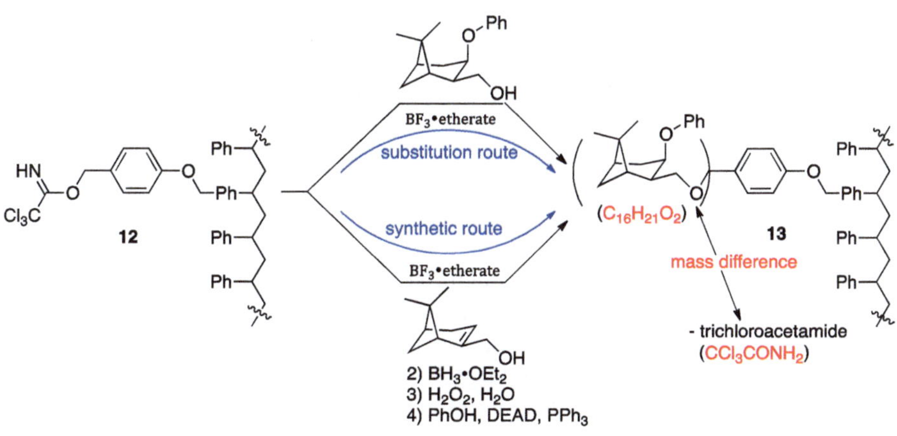

Scheme 4 Thought experiment demonstrating reaction stoichiometry on the solid-phase

Solid-Phase Synthesis Project **Resin Loading Calculator**

Enter value here	Date	7/29/16
Computed value		

What is the loading level of the commercial resin (mmol/g)? 1.06

(Polystyrene/DVB-Wang-trichloroimidate)

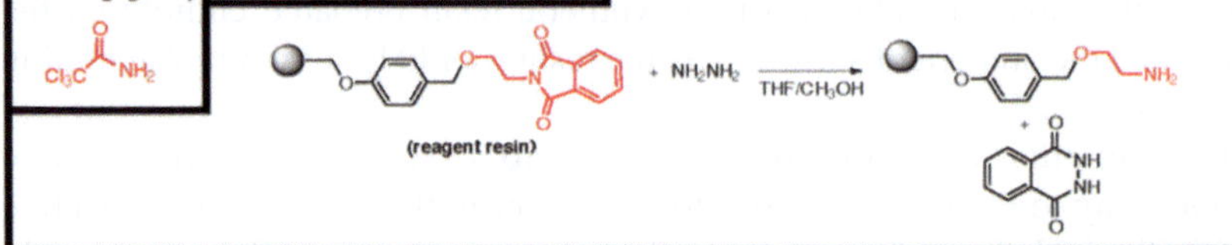

What is the mass of the leaving group on the commercial resin 161.5

(Cl, 35.5; OH, 17; trichloroimidate, 161.5)

What is the mass of the group attached in the reagent resin? 191

Reagent resin needed to have **1.06 mmol/gram** 1.03

=1+(F6*(F19-F16)/1000)

=1+(1.06*(191-161.5)/1000)

Mass (mg) of reagent resin used in reaction? 140

Number of mmol reacted 0.144

=((E24/1000)/F21)*F6

=((140/1000)/1.03)*1.06

Reagent 1:

How equivalents of reagent will you use? 4

What is the molecular wt of reagent? 32

Mass of reagent (mg) 18.4

Reagent 2:

How equivalents of reagent will you

What is the molecular wt of reagent

Mass of reagent (mg) 0.0

Fig. 7 Resin loading calculator spreadsheet

2-phthalimidoethanol compared to trichloroacetamide (see Scheme 5). Continuing with the synthesis, the next reagent resin would have 2-ethanolamine as the organic fragment. 2-ethanolamine weighs less than trichloroacetamide and the equivalent resin mass to have 1.06 mmol would be 0.93 g. For simplicity and consistency, one should always use the original leaving group as the standard mass for comparison (Scheme 5).

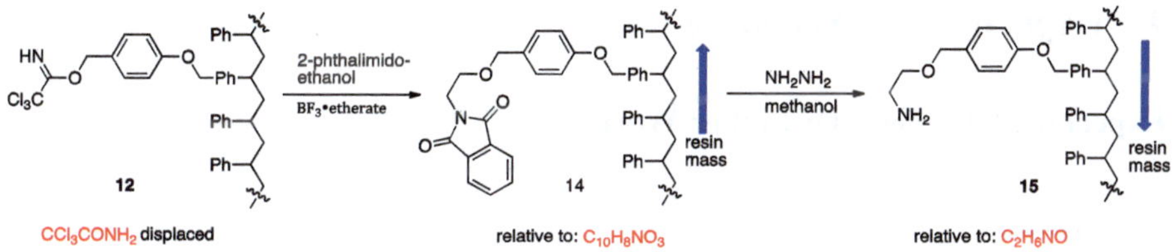

Scheme 5 Trichloroimidate (159 m/z) as the leaving group used throughout the synthetic scheme

The Resin Loading Calculator spreadsheet, shown in Fig. 7, provides a convenient way to set up stoichiometric calculations for solid-phase reactions using trichloroimidate as the leaving group. Three quantities must be entered into the highlighted open cells to initiate the calculation: (1) the loading level of the commercial resin, (2) the mass of the original leaving group on the commercial resin, and (3) the mass of the organic fragment on the reagent resin. The design of the actual experiment will require the mass of the resin being reacted, the number of chemical equivalence being used, and the molecular mass of the chemical reagent.

Students involved in this project used this calculator for all chemical reactions used in the indazole 1-oxide synthesis.

3 Laboratory Project

Synthetic compounds possessing the indazole 1-oxide or indazole core have become highly desirable compounds for biological studies. A recent review article by Cerecetto [25] itemized the diversity of biological activities exhibited by indazoles and benzindazoles. Newer investigations identified indazole-based compounds as potent agents with anti-inflammatory [26], anticancer [27], antimicrobial [28], antifungal [29, 30], antiangiogenic [31], and cytotoxic [32] activities. Viktor Krchnak is currently extending synthetic routes leading to the preparation of other nitrogenous heterocycles including quinazolines [33], and we have supplied combinatorial libraries of indazole 1-oxides to facilitate his research.

In this project we repeated and expanded the library of indazole 1-oxides published in the *Journal of Organic Chemistry* [9]. All students will start with Wang PS-resin and, depending on your assigned diversity tree, you will synthesize one of ten different indazole 1-oxides. As you proceed, you will need to document your work in both your laboratory notebook and in an interim report and a final report.

4 Experimental Schedule

Experiment 1 (Week 1): Linker Work

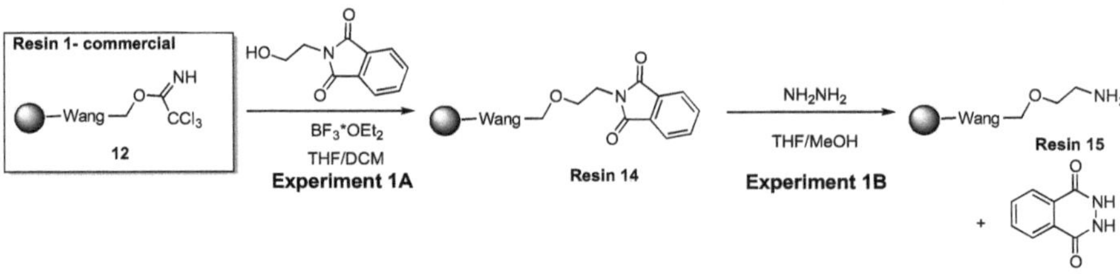

Experiment 1A: Trichloroimidate Replacement: (wear gloves) The method of Hannessian [24] for the substitution of trichloroimidate resin was followed. To Wang trichloroimidate resin (**Resin 12**) (150 mg pre-weighed into 5 mL filter syringe, 0.85 mmol/g) was added anhydrous THF (3 mL) which swelled the polystyrene resin (method: Video 1). After slurried for 0.2 h the THF is pushed out of the syringe and the filtered resin was ready for reaction. In a 5 dram vial was prepared a solution of *N*-(2-hydroxyethyl)phthalimide (49 mg, 0.26 mmol) in anhydrous THF (2.0 mL). To this new slurry was added BF_3·etherate (2.0 mL, freshly prepared stock solution as 0.010 M BF_3·etherate in anhydrous DCM). After capping the syringe, the entire assembly was placed on the rotisserie at RT for 0.5 h. The synthetic resin was filtered in the syringe and was washed consecutively with THF (3 × 3 mL) and DCM (3 × 3 mL) on the Domino Block synthesizer (method: Video 2). The resin was dried in a nitrogen stream for 0.3 h (method Video 3) (Table 1).

Experiment 1A Analysis: Pull open the filter syringe and weigh ~10 mg of the dried resin 2 into a tall vial and seal the vial. (This sample can be stored up to 1 week.) Tare an agate mortar on an analytical balance and then pour your reserved resin 2 sample (8–10 mg) into the mortar. Next add dry KBr to the mortar until the total mass is between 65 and 70 mg. Grind the solids together with the agate pestle until there is a consistency of very fine powder. By mass difference, scrape about 50–55 mg of the solid into a KBr die press (reweigh the mortar and add more powder to the die press until only 15 mg remains in the mortar). Put the second bolt into the die press and tighten. Tap the die press on the bench periodically as you finger tighten the bolts. Use the torque wrench to tighten to 35 ft./lbs. Immediately loosen the bolts and remove both bolts. You should be able to see a transparent pellet in the die. Two bands associated with trichloroimidate (3,339 (m), 1,661 (s) cm^{-1}) should be missing, and a strong band at 1,670 cm^{-1} should be present.

*<**Note:** Experiment 1B can be assigned for lab day 1 followed the next day by resin isolation, or the first part of Experiment 1B can be assigned the day before the next lab session. Staffing and safety considerations will determine which way to assign this experiment.>*

Table 1 Indazole 1-oxide targets to be prepared by students

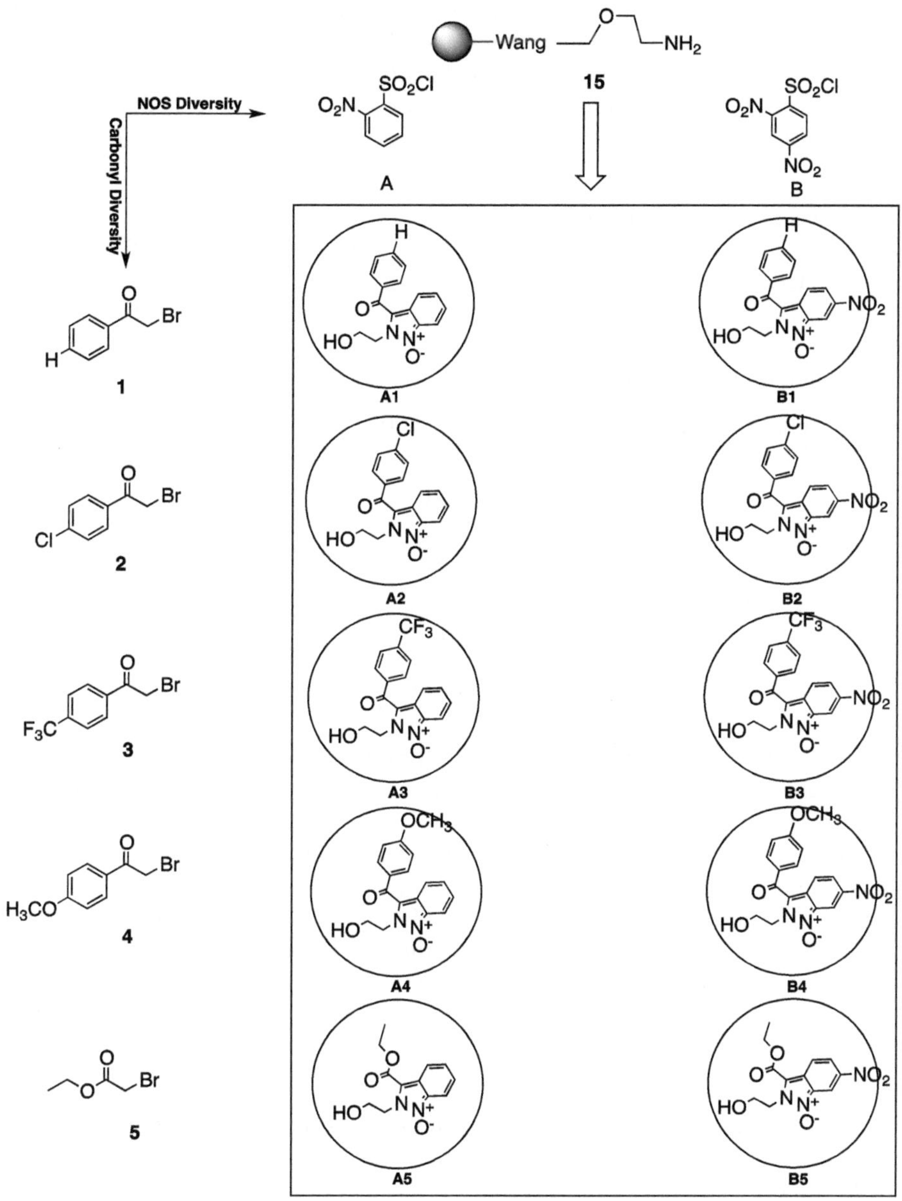

Experiment 1B: Imide Deprotection: (wear gloves) **Resin 14** (~140 mg) was swollen with a 1:1 mixture of THF:methanol (3 mL) (method: Video 1). After slurried for 0.2 h the THF:methanol is pushed out of the syringe and the filtered resin was ready for reaction with the hydrazine solution by pulling the hydrazine solution into the syringe (3 mL, prepared stock solution as 1.0 M hydrazine in 1:1 atmospheric THF:methanol). The filter syringe was capped and mixture tumbled overnight at RT.

Day 2- Experiment 1B (Continuation)

The reactive solution was removed from the reaction syringe and the filtered resin was washed consecutively with DMSO (2 × 3 mL), THF (3 × 3 mL), and DCM (3 × 3 mL) on the Domino Block synthesizer (method: Video 2). The resin was dried in a nitrogen stream for 0.3 h (method: Video 3). Recap the syringe and place the syringe in the jar desiccator found in the refrigerator.

Experiment 2 (Week 2): Introduction of NOS

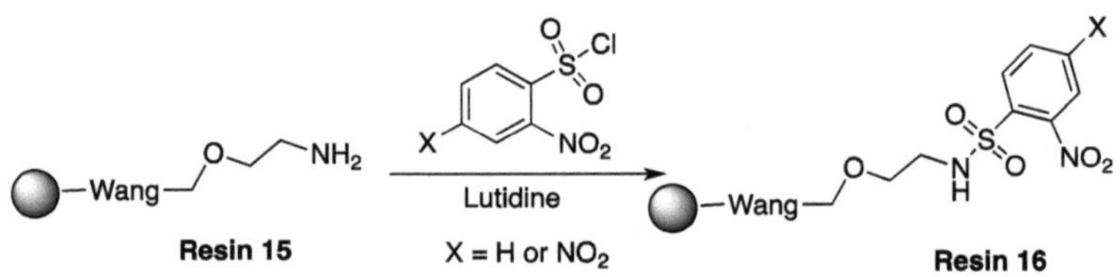

Experiment 2A

Experiment 2A: Preparation of Sulfonamide Resin

Note: Always allow syringes to warm to room temperature prior to opening containers in which they are stored.

The remaining **resin 15** (~0.14 g) in the 5 mL filter syringe was washed with DCM (2 mL) for 0.1 h and filtered. Into a 5 dram vial was weighed either 2-Nos-Cl (49 mg, 0.22 mmol) or 4-nitro-2-NOS-Cl (59 mg, 0.22 mmol) (based on original chemical assignment) and then 0.11 M lutidine (in DCM (3 mL) was added to the vial and the solid was dissolved. Once the mixture was homogeneous pull the entire solution into the 5 mL filter syringe. The slurry was placed on the rotisserie overnight (<24 h) at ambient temperature.

Day 2- Experiment 2A

The filter syringe is attached to the Domino Block synthesizer (method: Video 2) and the synthetic resin (2) is washed 3 times with DMF and 3 times with DCM on the Domino Block synthesizer and then dried with nitrogen (method: Video 3). Cap the syringe and store resin 3 in the jar desiccator in the refrigerator.

Experiment 3 (Week 3): Indazole 1-Oxide Formation and Analysis

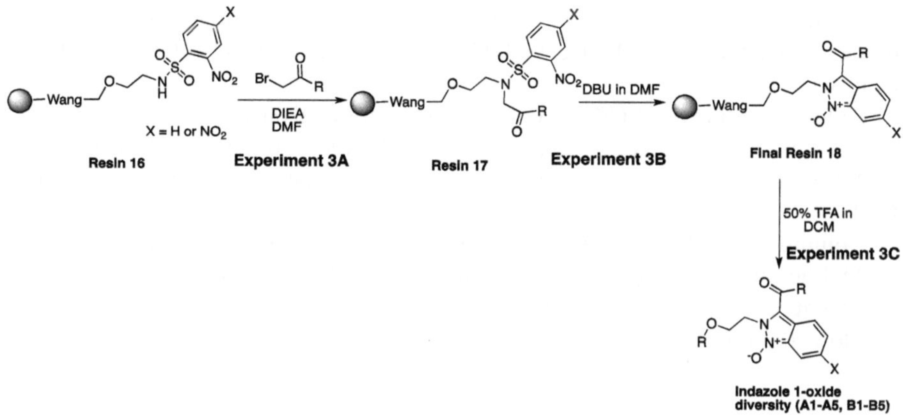

Day Before Laboratory Section

Experiment 3A: Alkylation of Resin 16 with Bromoketone/Bromoester

Note: Prior experience with the Domino Block synthesizer and with reaction set up has prepared you to do this synthetic step independently. Stockroom personnel and safety officers are in the building for your safety and security.

Note: Warm the syringe and sample to room temperature before washing the resin.

Resin **16**, ~150 mg in the filter syringe, was washed 3 times with DCM and 5 times with DMF on the Domino Block synthesizer (do not dry the resin!). A solution of 0.5 M bromoketone/bromoester (diversity A–D) (in DMF (2 mL)) and diisopylethylamine (3 mmol, 522 μL, automatic pipette) was added into a 10 dram vial and the combined solution was pulled into the filter syringe containing swollen resin 4. The slurry was placed on the rotisserie overnight (<24 h) at ambient temperature.

Experiment 3B: Preparation of Indazole 1-Oxide Resins

Resin 17, ~150 mg in the syringe, was washed 5 times with DMF (anhydrous DMF bottle). A solution of 0.2 M DBU in DMF (2 mL) was added to the resin in the filter syringe, and the resin slurry was mixed on the rotisserie at ambient temperature for 30 min. Note color changes. The resin, **6**, was filtered and washed 3 times with DMF, 5 times with DCM, and 3 times with MeOH, and dried with nitrogen.

Experiment 3C: Indazole Oxide Cleavage and LC/MS Analysis

Analytical samples of indazole 1-oxides, (**A1–A5, B1–B5**), were obtained after cleavage with 50% TFA in DCM for 1 h at room temperature (see video 4). Approximately 10 mg of resin **18** was weighed into a tall 1 mL vial and 8 drops of 1:1 TFA/DCM solution was added. The mixture was agitated on a shaking titer plate for 1 h. The mixture was dried in a nitrogen stream and residue was suspended in acetonitrile (0.5 mL), and after thorough mixing the suspension is pulled into a

1 mL syringe. An HPLC filter was placed on the syringe and the acetonitrile extracts are collected into an HPLC vial. The resin was rinsed a second time with acetonitrile and the HPLC filter was reattached to the syringe and the second extracted is added to the HPLC vial. Cap and seal the vial and submit for LC/MS analysis.

Notes regarding the LC/MS conditions used for analysis:

LC = Waters Alliance Separations Module (Waters, Corporation, Milford, MA).

MS = Thermo-Fisher LXQ linear ion trap mass spectrophotometer (San Jose, CA).

Column: YMC-Pack Pro C18, 3 × 50 mm, 3 μm, 12 nm pore size.

Mobile phase: A = 10 mM ammonium acetate; B = acetonitrile.

Eluent mix: 95% A to 20% A over 15 min.

Flow: 0.3 mL/min

5 Summary

This chapter presents an instructional module developed for a second-year undergraduate Organic Chemistry laboratory. Through the use of a traditional laboratory manual, training videos (4 videos, 2–3 min in length) highlighting four recurring manipulations in semi-automated solid-phase synthesis and hands-on learning experience, students gained sufficient operational skill to work independently with this equipment. While interesting, the synthesis of indazole 1-oxides is a surrogate chemical class to illustrate that a wide variety of structural motifs can be synthesized. Additionally, this indazole 1-oxide synthesis module should serve as a benchmarking tool to help laboratory instructors set workload expectations for the students.

Educational outcomes from this module are not limited to in-lab activities. Because of the repetitive nature of solid-phase synthesis, dexterity and manipulation of the equipment is quickly mastered. The speed of solid-phase reactions also allows students to conduct related reactions and/or characteristic reactions of a particular functional group. For instance, the indazole 1-oxide synthesis in this report uses amine deprotection, amine sulfonylation, and amine (sulfonamide) alkylation in sequence. All of these reactions are discussed in a textbook, but the impact on student learning is quite different. The ability to design syntheses that build and compound repetitively on one or two functional groups is a powerful tool for student learning.

From a cognitive perspective, a 1-week delay between repetitive yet similar activities is suited for long-term retention of knowledge (storage strength) [34–36]. One notable outcome from this laboratory module is the enhanced understanding of reaction kinetics and thermodynamics. The pervasiveness of enhanced learning became evident on formal evaluative instruments like examinations and the American Chemical Society national field exam. Phenomenological, the final cyclization step, the step that forms indazole 1-oxide, is evident from an intense color change in the resin due to changes in conjugation. The diversomers in this project have somewhat different rate constants for indazole 1-oxide cyclization, and this can be observed as the reaction proceeds. As such, students gained a qualitative sense for how chemical structure influences chemical rates by observing how fast different diversomers change color. After this project, concepts like "Le Chatelier's Principle" and concentration effects are no longer about pointing left or right for a chemical equation. Students manifest newfound understandings of equilibria and reaction dynamics to unrelated chemical systems.

Finally, the evidence presented in this chapter should persuade readers that solid-phase synthesis can be mastered quickly by general practitioners of organic chemistry. Equipment used herein such as the Domino Block synthesizer is not fully necessary to get started; however, this device and others used here enhance the user experience. Polypropylene filter syringes, needles, a simple aspiration system and appropriate polystyrene resin is all that is needed to start a chemical synthesis.

Acknowledgment This work was supported by the Chemistry Department, Kalamazoo College; and the author wishes to thank Dr. Laura Furge, Chemistry Department, Kalamazoo College for her work in the instructional laboratories; and additional support for this project in part by a grant to Kalamazoo College from the Howard Hughes Medical Institute through the Precollege and Undergraduate Science Education Program.

References

1. Krchnak V, Vagner J (1990) Peptide Res 182
2. Krchnak V, Padera V (1998) Bioorg Med Chem Lett 3261
3. Fuller AA (2016) J Chem Ed 953
4. Scott WL, Denton RE, Marrs KA, Durrant JD, Samaritoni G, Milata MA, Brown SP, Carnahan JM, Fischer LG, Glos CE, Sempsrott PJ, O'Donell MJ (2015) J Chem Ed 819
5. Pohl NLB, Kirshenbaum K, Yoo B, Schulz N, Zea CJ, Streff JM, Schwarz KL (2011) J Chem Ed 999
6. Seymore E, Hunter A-B, Laursen SL, Deantoni T (2004) Sci Educ Sci 493
7. Anderson WA, Banerjee U, Drennan C, Elgin SCR, Epstein IR, Handelsman J, Hatfull GF, Losick R, O'Dowd D, Olivera BM, Strobel SA, Walker GC, Warner IM (2011) Science 152
8. Olsen S, Riordan DG (2012) Engage to Excel: Producing One Million Additional College Graduates with Degrees in Science, Technology Engineering, and Mathematics. Report to the President Executive Office of the President
9. National Research Council (2012) Discipline-Based Education Research: Understanding and Improving Learning in Undergraduate Science and Engineering. In: Singer SR, ielsen NR, chweingruber HA (eds) Committee on the Status, Contributions, and Future Directions of Discipline-Based Education Research. Board on Science Education, Division of Behavioral and Social Sciences and Education. National Academies Press, Washington
10. Merrifield RB (1963) J Am Chem Soc 2149
11. Merrifield RB, Stewart JM (1965) Nature 522
12. Merrifield RB (1968) Sci Am 56
13. Merrifield RB (1964) Biochem 1385
14. Marglin A, Merrifield RB (1966) J Am Chem Soc 5051
15. Takashima H, Du Vigneaud V, Merrifield RB J Am Chem Soc 1323

16. Barany G, Merrifield RB (1979) The peptides, vol 2. Academic, New York, pp 1–284
17. Malanga M, Isogai O, Yamada T, Iwasaki S, Kuramoto M 2010 In: Schellenberg J (ed) Syndiotactic polystyrene, 1st edn. Wiley, Hoboken, pp 3–13
18. AAPPtec website: https://www.aapptec.com/core-resins-i-250.html?osCsid=c3mnaji4r1suekr7c4i5keukv3
19. Merrifield RB, Littau V (1968) In: Bricas E (ed) Peptides 1968. North-Holland Publishing Co., Amsterdam, p 179
20. Hodge P (1997) Chem Soc Rev 417
21. Marchetto R, Cilli EM, Jubilut GN, Schreier S, Nakaie CR (2005) J Org Chem 4561
22. Bouillon I, Soural M, Miller MJ, Krchnak V (2009) J Comb Chem 213
23. Wang S-S (1973) J Am Chem Soc 1328
24. Hanessian S, Xie F (1998) Tetrahedron Lett 733
25. Cerecetto H, Gerpe A, Gonzalez M, Aran VJ, de Ocariz CO (2005) Mini-Rev Med Chem 869
26. Chen CJ, Hsu MH, Huang LJ, Yamori T, Chung FG, Lee FY, Teng CM, Kuo SC (2008) Biochem Pharmacol 360
27. Yakaiah T, Lingaiah BPV, Narsaiah B, Shireesha B, Kumar BA, Gururaj S, Parthasarathy T, Sridhar B (2007) Bioorg Med Chem Lett 3445
28. Yakaiah T, Lingaiah BPV, Narsaiah B, Kumar KP, Murthy USN (2008) Eur J Med Chem 341
29. Park JS, Yu KA, Yoon YS, Han MR, Kang TH, Kim S, Kim NJ, Yun H, Suh YG (2007) Drugs Future 121
30. Park JS, Yu KA, Kang TH, Kim SH, Suh YG (2007) Bioorg Med Chem Lett 3486
31. Huang LJ, Shih ML, Chen HS, Pan SL, Teng CM, Lee FY, Kuo SC (2006) Bioorg Med Chem 528
32. Rakib E, Oulemda B, Abouricha S, Bouissane L, Mouse HA, Zyad A (2007) Lett Drug Des Discovery 467
33. Krupkova S, Slough GA, Krchnak V (2010) J Org Chem 4562
34. Kober N (2015) Reaching students: what research says about effective instruction in Undergraduate Science and Engineering. Board on Science Education, Division of Behavioral and Social Sciences and Education. The National Academies Press, Washington
35. Carey B (2014) How we learn. Penguin Random House, New York, p. 82
36. Schmidt RA, Bjork RA (1992) Psych Sci 207

Top Heterocycl Chem (2017) 52: 23–64
DOI: 10.1007/7081_2016_2

Published online: 13 April 2017

Solid-Phase Synthesis of Seven-Membered Heterocycles with Two Nitrogen Atoms

Veronika Fülöpová and Miroslav Soural

Abstract Due to the importance of diazepines and diazepanes in medicinal chemistry and chemical biology, their preparation from diverse starting materials has been frequently reported. In this chapter, we summarize all strategies employing the method of solid-phase synthesis. More than seventeen different types of target compounds are accessible. The individual approaches are grouped according to the type of the target scaffold.

Keywords Benzodiazepines • Combinatorial chemistry • Diazepanes • Diazepines • Solid-phase synthesis

Contents

V. Fülöpová and M. Soural (✉)
Department of Organic Chemistry, Institute of Molecular and Translational Medicine, Faculty of Science, Palacky University, 17. listopadu 12, Olomouc 771 46, Czech Republic
e-mail: miroslav.soural@upol.cz

Abbreviations

AA	Amino acid
Boc	*t*-Butyloxycarbonyl
BAL	Backbone amide linker
BTC	Bis(trichloromethyl) carbonate
DABCO	1,4-Diazabicyclo[2.2.2]octane
DBU	1,8-Diazabicyclo[5.4.0]undec-7-ene
DCC	*N*,*N*′-Dicyclohexylcarbodiimide
DCE	Dichloroethane
DCM	Dichloromethane
DIAD	Diisopropyl azodicarboxylate
DIBAL-H	Diisobutylaluminium hydride
DIC	*N*,*N*′-Diisopropylcarbodiimide
DIEA	*N*,*N*-Diisopropylethylamine
DEAD	Diethyl azodicarboxylate
DECP	Diethyl cyanophosphonate
DMAP	4-Dimethylaminopyridine
DMF	*N*,*N*-Dimethylformamide
DMSO	Dimethyl sulfoxide
DMTMM	4-(4,6-Dimethoxy-1,3,5-triazin-2-yl)-4-methylmorpholinium chloride
EEDQ	*N*-ethoxycarbonyl-2-ethoxy-1,2-dihydroquinoline
Fmoc	Fluorenylmethoxycarbonyl
HATU	[Bis(dimethylamino)methylene]-1*H*-1,2,3-triazolo[4,5-*b*]pyridinium 3-oxid hexafluorophosphate)
HOBt	1-Hydroxybenzotriazole
MBHA	Methylbenzhydrylamine
NMP	*N*-Methylpyrrolidone
Nos	Nitrobenzenesulfonyl
TBTU	*N*,*N*,*N*′,*N*′-Tetramethyl-*O*-(benzotriazol-1-yl)uronium tetrafluoroborate
TBP	Tributylphosphine
TEA	Triethylamine
THF	Tetrahydrofuran

TFA	Trifluoroacetic acid
TFMSA	Trifluoromethanesulfonic acid
TPP	Triphenylphosphine

1 Introduction

Within the large family of heterocyclic compounds, diazepines and diazepanes are an interesting group of derivatives with specific physical-chemical and biological properties. Depending on the position of the two nitrogen atoms in the seven-membered scaffold, three general classes of compounds are distinguished, namely, 1,2-diazepanes, 1,3-diazepanes, and 1,4-diazepanes, as well as the corresponding unsaturated analogues, the diazepines (Fig. 1).

From a historical perspective, the most frequently studied class in this field of research is the [1,4]diazepines, which are typically condensed with a benzene ring. Although the majority of these compounds reported to date are of synthetic origin, the scaffold is also found in natural products such as diazepinomicin [1] or callysponine [2]. Synthetic benzo[1,4]diazepines have been identified as important molecules in drug discovery, especially as potent central nervous system (CNS) agents with strong antianxiety, muscle relaxant, and tranquilizing effects due to their specific binding sites within $GABA_A$ receptors [3]. Although the CNS effect of benzodiazepines is considered most important, there are many other beneficial properties, such as an antiarrhythmic effect [4], antagonism of cholecystokinin receptors [5], inhibition of HIV-1 reverse transcriptase [6], opioid receptor activity [7], anticancer [8] or anti-inflammatory [9] effects, and many others. For these reasons, benzodiazepines are conventionally classified as "privileged scaffolds".

The gold era of the solid-phase syntheses of heterocycles began in the mid-1990s. Among other methods to prepare heterocyclic scaffolds, a variety of methodologies to prepare diazepane/diazepine derivatives was also developed. Unsurprisingly, the medicinal importance of benzo[1,4]diazepines reported by Sternbach [10] redirected the research focus of solid-phase chemists so that the majority of reported results was related to this group of compounds. Interestingly, the first literature evidence of benzo[1,4]diazepine solid-phase synthesis occurred in 1977 (www.espatentes.com/pdf/0445831_A1.pdf), only twelve years after

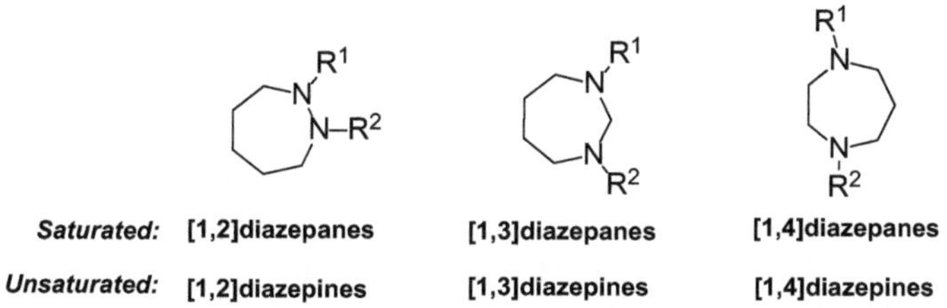

Fig. 1 General structural classification of seven-membered cycles with two nitrogen atoms

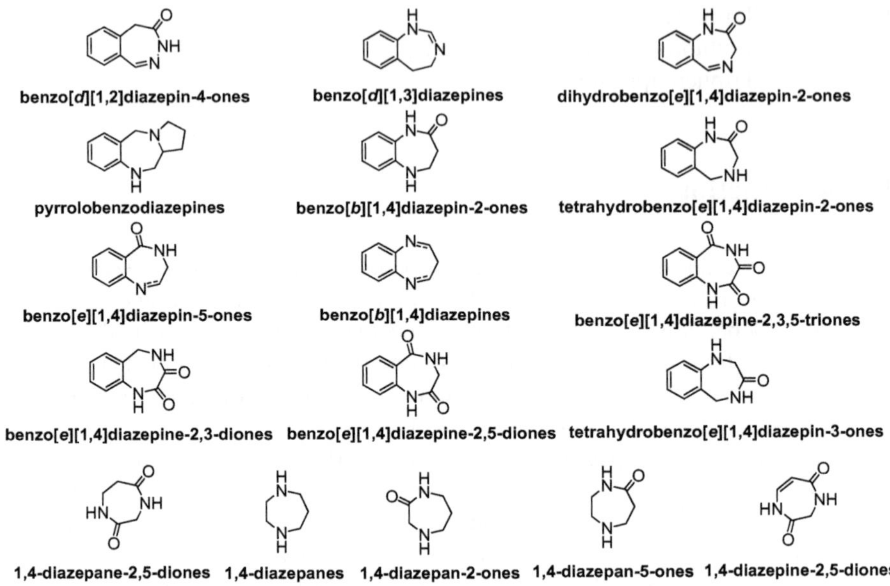

Fig. 2 Individual scaffolds reported using solid-phase synthesis

Merrifield's introduction of solid-phase peptide synthesis [11] and almost twenty years before the beginning of the solid-phase heterocyclic era.

In 2006, the preparation of benzodiazepines was reviewed by Kamal [12] and was partially addressed in several other review articles [13, 14]. The following section summarizes all strategies reported to date. Although a significant part of the text is devoted to benzodiazepine chemistry, the methods are extended to the entire group of diazepane/diazepine derivatives. The individual approaches are grouped according to the type of scaffold, as depicted in Fig. 1, and are divided into subsections devoted to more specific scaffolds (Fig. 2).

2 1,2-Diazepines

2.1 Benzo[d][1,2]diazepin-4-ones

To date, there is a single strategy applicable for the solid-phase synthesis of this scaffold containing compounds reported by Bevacqua et al. [15] This approach is based on using aminomethyl polystyrene resin **1** coupled with 4-(hydroxymethyl) benzoic acid. The resin was acylated with 3,4-dimethoxyphenylacetic acid, followed by the Friedel-Crafts acylation with different benzoyl chlorides to yield resin-bound **4**. The final cleavage with hydrazine afforded the target 1-aryl-3,5-dihydro-4*H*-2,3-benzodiazepin-4-ones **5** (Scheme 1) as compounds with potentially

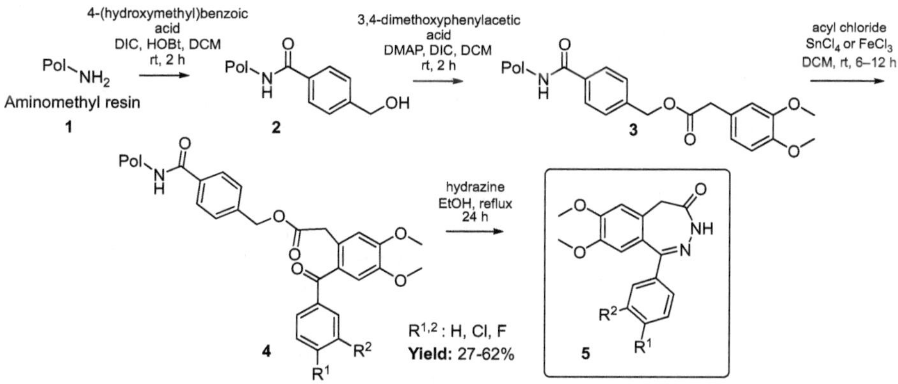

Scheme 1 Solid-phase synthesis of 2,3-benzodiazepin-4-ones

antiepileptic properties. The authors also attempted to synthesize 1-alkyl analogues of products **4**, successfully performing acylation with propionyl chloride. However, the cyclization with hydrazine did not give the desired 1-ethyl-2,3-benzodiazepin-4-ones.

3 1,3-Diazepines

3.1 Benzo[d][1,3]diazepines

The preparation of 1,3-benzodiazepines as novel dopamine antagonists was described by Zhu et al. [16]. The phenolic building block **6** was attached to a brominated Wang resin, and after deprotection and acetylation of the amino group, the nitro derivative **9** was reduced to aniline **10** with tin(II)chloride dihydrate. The intermediate **10** was reductively alkylated and cleaved from the resin using trifluoroacetic acid (TFA). Cyclization of the final diazepines **16a** was accomplished in the solution phase using $POCl_3$ (Scheme 2). Similarly, solid-phase *N*-arylation chemistry using boronic acids and $Cu(OAc)_2$ was used to generate N^1-arylbenzodiazepines [17]. Alternatively, intermediate **8** was reduced and cyclized with thiophosgene to give the cyclic thiourea **13**. Subsequent methylation, amination, and cleavage led to 2-amino-1,3-benzodiazepines **16b**.

In 2011, Yu et al. [18] reported the solid-phase synthesis of similar compounds based on a different approach. Methylbenzhydrylamine (MBHA) resin **17a** was acylated with diverse Boc-L-amino acids. After the protecting group was removed, resin **18** was acylated with 2-bromophenyl acetic acids, followed by the reduction of diamide **19** with BH_3-THF. Cyclization with cyanogen bromide gave the monocyclic guanidine **21**. Target benzodiazepine derivatives **22** were obtained by treatment with $Pd(OAc)_2$, $Cu(OAc)_2$, and Cs_2CO_3 in dimethylformamide (DMF)

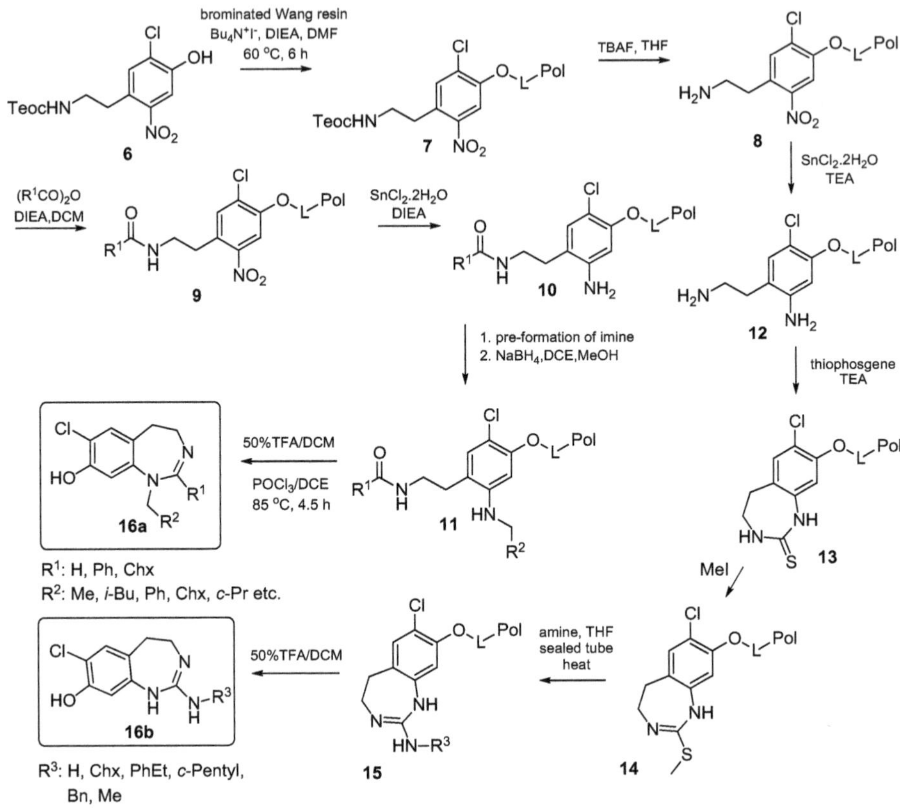

Scheme 2 Solid-phase synthesis of 1,3-benzodiazepines

(Scheme 3). The reaction did not require any ligands or additives and was performed under air.

3.2 [1,3]Diazepines

The only reported solid-phase strategy to prepare 1,3-diazepine derivatives is based on the Baylis-Hillman reaction [19]. Wang resin **24** acylated with acryloyl chloride was subjected to reaction with aldehydes in the presence of 1,4-diazabicyclo[2.2.2] octane (DABCO). Baylis-Hillman adducts **25** were then converted to acetates **26** using acetyl chloride in pyridine. Subsequently, a Michael addition with 1,4-diaminobutane resulted in allyl amine derivatives **27**. After cyclization of the diazepine scaffold with cyanogen bromide, the cyclative cleavage induced with triethylamine (TEA) was performed to release the final products **29** from the resin (Scheme 4). NMR studies revealed the exclusive formation of E-isomers.

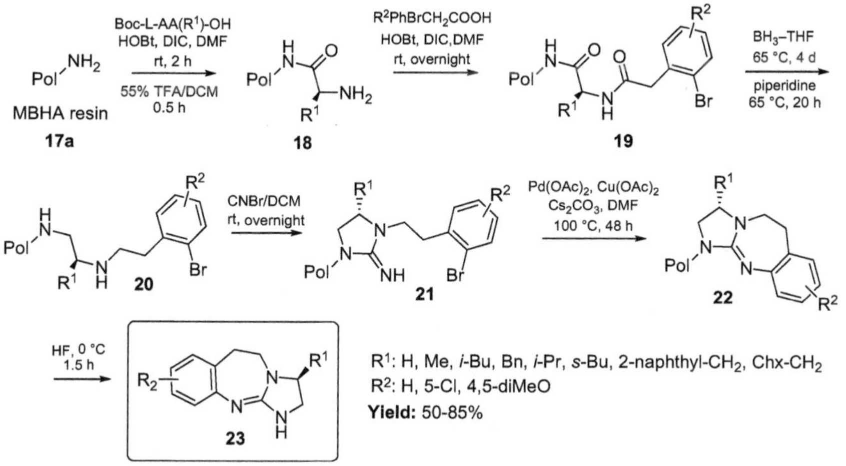

Scheme 3 Solid-phase synthesis of 1,3-benzodiazepines via a Pd-catalyzed intramolecular coupling reaction

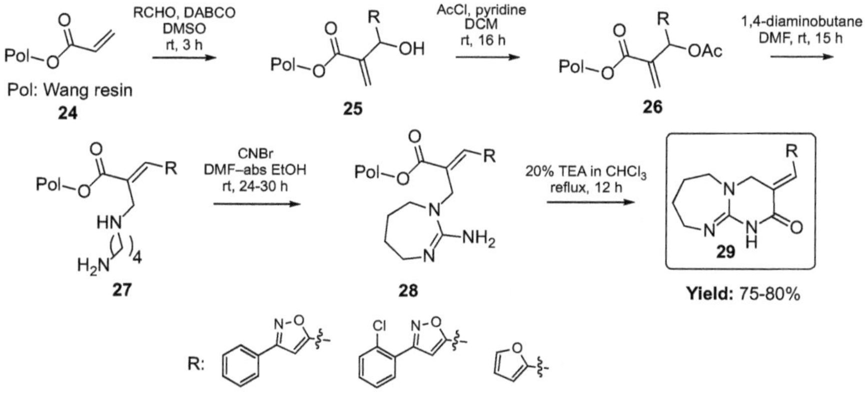

Scheme 4 Solid-phase synthesis of 1,3-diazepines based on the Baylis-Hillman reaction

4 1,4-Diazepines

4.1 Dihydro-benzo[e][1,4]diazepin-2-ones

As previously mentioned, 1,4-benzodiazepines are the most frequently reported diazepines in solid-phase synthesis. The first contributions appeared in the early 1990s and were devoted to the preparation of 1,3-dihydro-2*H*-benzo[*e*][1,4] diazepin-2-ones, which are related to the structure of CNS benzodiazepine modulators. The pioneer in this field was Jonathan Ellman, who reported a solid-phase synthesis starting from 2-aminobenzophenones **30** attached to a polystyrene solid support through either a hydroxy or carboxylic acid functionality employing the

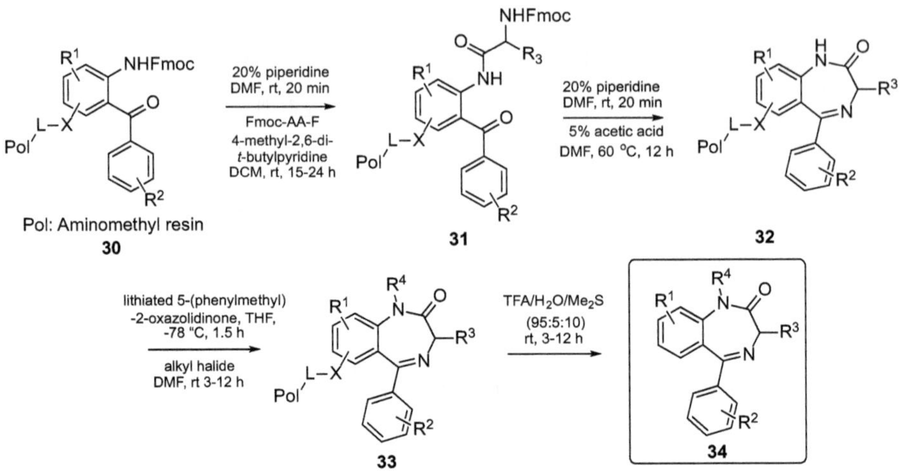

L: [4-(hydroxymethyl)phenoxy]acetic acid
R^1: Cl, COOH
R^2: H, OH
R^3: Me, *i*-Pr, CH_2COOH, $(CH_2)_4NH_2$, Bn, 4-MeO-Bn
R^4: H, Me, *i*-Pr, Bn
X: O, COO
Yield: 85-100%

Scheme 5 Synthesis of 1,4-benzodiazepines by Ellman

acid-cleavable linker [4-(hydroxymethyl)phenoxy]acetic acid [20]. After removal of the Fmoc-protecting group and acylation with Fmoc-amino acids, the ring closure to **32** was performed by heating the unprotected resin in diluted acetic acid. The fourth diversity position R^4 was created using alkyl halides, followed by the cleavage of final compounds **34** from the resin in very high yields (Scheme 5). Two years later, Ellman employed his strategy for the combinatorial synthesis of 192 structurally diverse benzodiazepines using Geysen's pin apparatus [21].

Despite the great diversity in the structure of the final compounds **34**, the disadvantage of this method from a combinatorial perspective is the limited availability of the starting compounds, 2-aminoaryl ketones, which must be pre-synthesized in the solution phase. To improve his initial method, Ellman developed a strategy applicable to the preparation of these building blocks directly on the solid phase [22]. The key intermediate **36** was synthesized from *p*-aminophenol **35** in five steps and was attached to an aminomethyl polystyrene resin (Scheme 6). The resin **37** was then subjected to Stille coupling with aroyl- or alkyl chlorides. To minimize protodestannylation and premature carbamate deprotection, potassium carbonate and diisopropylethylamine (DIEA) were used as acid scavengers. The modification of intermediate **39** was continued according to the previously reported protocols (see Scheme 5). The developed strategy provides rapid access to a large number of diverse 2-aminoaryl ketone derivatives **39** from commercially available acid chlorides.

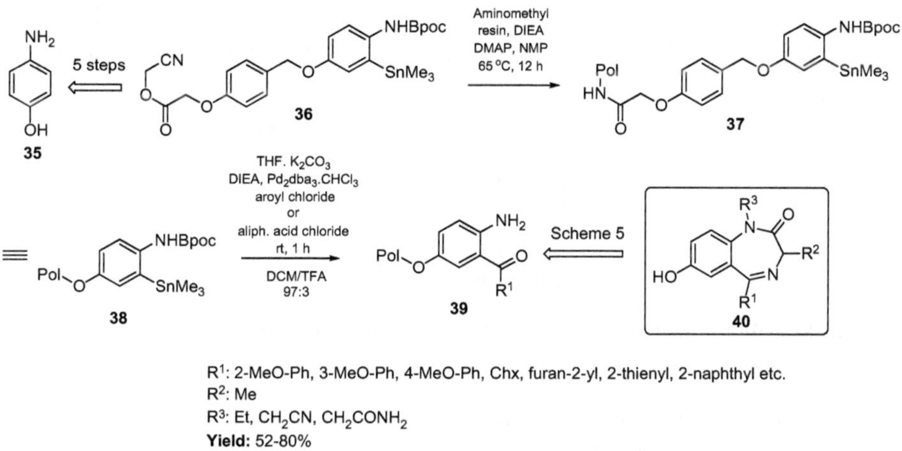

Scheme 6 Preparation of 1,4-benzodiazepines via the solid-phase synthesis of aminoketone building blocks

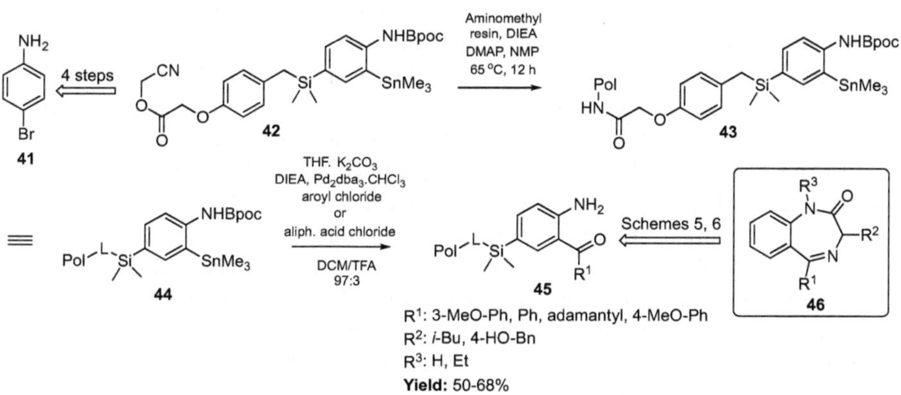

Scheme 7 Traceless synthesis of 1,4-benzodiazepines

Ellman further reported the preparation of 1,4-benzodiazepines in a traceless manner [23]. The incentive for the development of this alternative approach was that after cleavage from the solid support at the end of a synthesis sequence, the functional group resulting from the linker body (e.g., a hydroxy group, see Scheme 6) can have a negligible, positive or negative effect on the biological or chemical activity of the target molecule, depending on where it is situated. Therefore, silicon-based linker **42** was synthesized from 4-bromaniline **41** in four steps (Scheme 7). After attaching the linker to the aminomethyl resin and completing the reaction sequence according to Schemes 5 and 6, the final compounds **46** were cleaved with either HF or TFA, leaving behind no trace or "memory" of the solid-phase synthesis. Later, silicon was replaced with germanium, resulting in an easy electrophilic

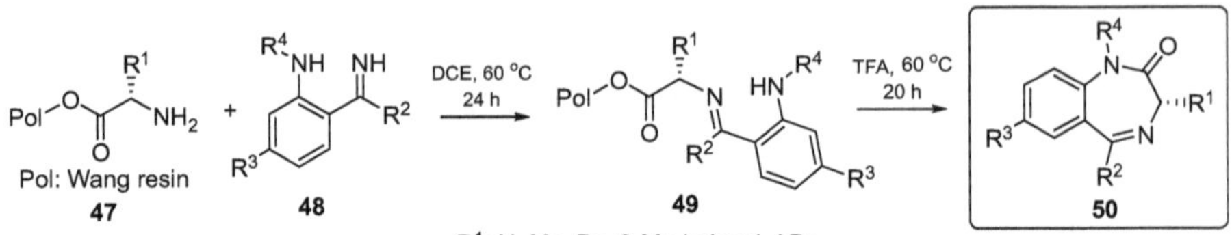

R¹: H, Me, Bz, 3-Me-indanyl, *i*-Pr
R²: Ph, 4-MeO-Ph, Chx, 2-Thn
R³: H, Cl, NO_2
R⁴: H, Me
Yield: 9-56%

Scheme 8 DeWitt's method for the solid-phase synthesis of 1,4-benzodiazepines

demetalation, which significantly increased the tolerance level of the sequence toward diverse functional groups [24].

In 1993, DeWitt reported a three-step traceless synthesis of 1,4-benzodiazepines from polymer-supported amino acids **47** and 2-amino benzophenone imines **48** [25, 26]. To obtain the final compounds, resin **49** was treated with neat TFA at an elevated temperature, which caused the consecutive cleavage and cyclization to **50** (Scheme 8). Forty benzodiazepines were synthesized using this method and were isolated with a crude purity above 90%. A few years later, DeWitt's method was successfully utilized to introduce a novel polymer support derived from an ion exchange resin [27]. Furthermore, in 2007, Sams followed DeWitt's protocol employing a "catch and release" strategy [28]. To avoid the purification requirement of the crude imine building blocks **48**, a catch with polymer-supported amino acids **47** was applied, followed by washing the unreacted side products from the resin using *N*-methylpyrrolidone (NMP). To obtain the final benzodiazepines **50** in high purity, Sams modified the cleavage protocol and replaced TFA with acetic acid. In this way, the products were released from the resin only by the cyclative cleavage, thus avoiding the potential contamination of the final compounds with cleaved linear intermediates **49**.

Ellman's and DeWitt's approaches were combined by Lattmann in a project aimed at preparing 1,4-benzodiazepines for the cholecystokinin (CCK) radioligand binding assay [29]. The Wang resin **51** was acylated with various Fmoc-amino acids that upon reaction with aminoketones gave the imine resin **53**. The acid-mediated cleavage resulted in the final products **55** (Scheme 9). Alternatively, a cyclative cleavage method was applied using pyridine and 4-*N*,*N*-dimethylaminopyridine (DMAP) to trigger the reaction. The synthesis of 168 benzodiazepines was achieved from eleven ketones and fourteen amino acids in a combinatorial fashion employing Synphase crowns.

In 2010, the immobilization of aminoketone building blocks on the alkoxyamine linker **56** was reported [30]. In contrast to Lattmann's procedure, the starting polymer-supported amino acids were replaced with resin **57** (Scheme 10). The authors reported optimized procedures for *N*-alkylation and managed to increase the enantiomeric excess of the final products **59** to 99%

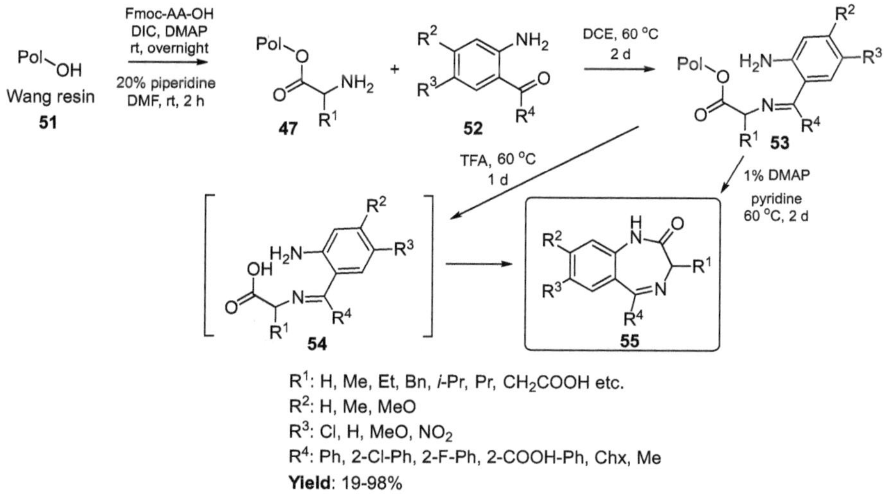

Scheme 9 Combined Ellman and DeWitt approaches reported by Lattmann

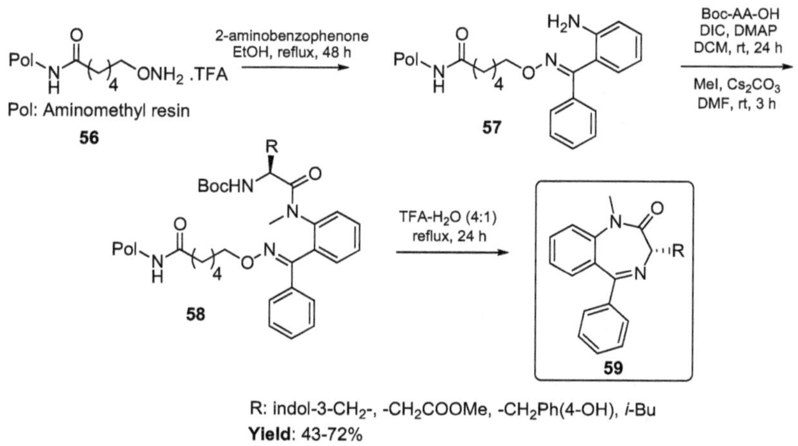

Scheme 10 Solid-phase synthesis of 1,4-benzodiazepines employing an alkoxyamine linker

using (1-[bis(dimethylamino)methylene]-1*H*-1,2,3-triazolo[4,5-*b*]pyridinium 3-oxid hexafluorophosphate (HATU) instead of DMAP in the acylation step.

The preparation of heterocyclic analogues of 1,4-benzodiazepines with a condensed indole instead of a benzene ring was reported by Lee [31]. The key intermediate **60** was synthesized from a polystyrene chlorotrityl resin and 2-chloro-3-formyl-5-hydroxyindole in four steps. After acylation with Fmoc-amino acids to give resin **61**, the Fmoc-cleavage was followed by spontaneous cyclization to yield the tricyclic products **62** (Scheme 11). Interestingly, the commonly used procedure employing a piperidine/DMF cocktail afforded the desired

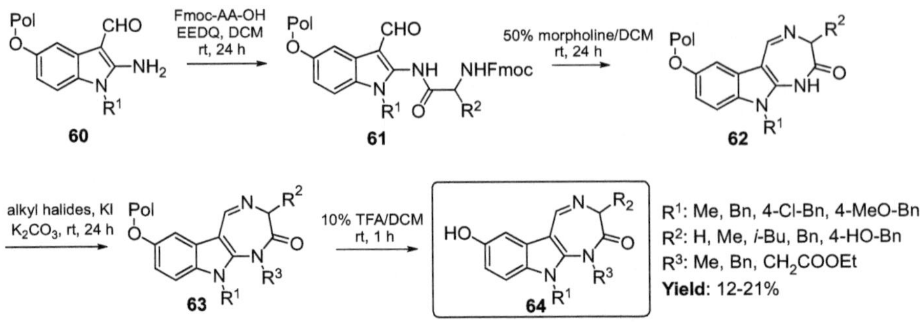

Scheme 11 Solid-phase synthesis of indole-diazepine-fused heterocycles

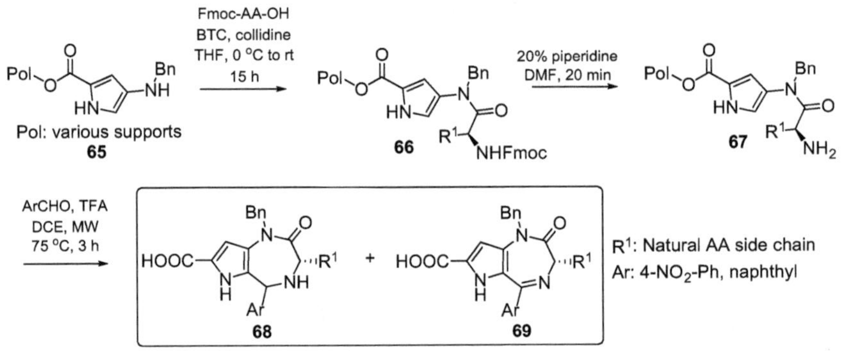

Scheme 12 Solid-phase synthesis of pyrrole-diazepine-fused heterocycles

products in very low yield. Therefore, the cleavage of the Fmoc-protecting group was performed with 50% morpholine/DCM. Prior to the cleavage, the N^1-alkylation was accomplished using alkyl halides.

In the field of research concerning dihydrobenzo[*e*][1,4]diazepin-2-one heterocyclic analogues, an interesting strategy for the preparation of compounds with a differential modulation of urotensin II receptor-mediated vasoconstriction was recently introduced by Lubell [32, 33]. As the key intermediate, resin-bound **65** was synthesized from immobilized 4-hydroxyproline. After acylation with Fmoc-amino acids and after Fmoc-protecting group cleavage, the final scaffold was obtained via the Pictet-Spengler cyclization using microwave heating with aldehydes (Scheme 12). The cyclization afforded a mixture of compounds **68** and **69** in variable ratios. The method was tested for different polymer supports (a Wang resin, a Merrifield resin, and a soluble TAP methylbiphenyl support), and the highest efficacy was reported for the Wang resin with average overall yields of 32%.

4.2 *Tetrahydrobenzo[e][1,4]diazepin-2-ones*

In addition to 1,3-dihydro-2*H*-benzo[*e*][1,4]diazepin-2-ones, considerable attention was also given to the solid-phase syntheses of their tetrahydro-analogues. The first contribution in this field was reported in 1997 by Bhalay et al. [34]. This approach was based on the acylation of Wang resin **51** with fumaroyl chloride, followed by the reaction of resin **70** with alcohols pre-synthesized from methyl anthranilates in two steps. After mesylation and reaction with various amines to give the polymer-supported intermediates **72**, the final compounds were obtained via cleavage and concomitant cyclization triggered by sodium methanolate (Scheme 13). This method was used to prepare 120 benzodiazepinones **73** by the combinatorial synthesis of twenty-four intermediates **71** and five amines R^3NH_2.

A different strategy was reported by Lou et al. [35]. Starting with the Rink amide resin or the reductively aminated BAL resin **74**, the reaction sequence continued by acylation with 4-(bromomethyl)-3-nitrobenzoic acid. Nucleophilic substitution with various amino acid methyl esters was performed followed by reduction of the nitro compounds **76** with $SnCl_2$ and saponification of methyl esters **77**; then, the seven-membered ring was cyclized using the HOBt method (Scheme 14). Note that the direct cyclization of methyl ester **77** was not achieved, even under harsh reaction conditions. Prior to the acid-mediated cleavage from the resin, the third diversity position was introduced employing N^4-reductive alkylation with aldehydes. This method was also applicable to cyclic amino acid esters, such as proline methyl ester, to give structurally constrained tricyclic compounds. The same approach was later utilized by Ede to synthesize the benzodiazepines **80** on Synphase lanterns [36]. However, in contrast to Lou's procedure, the N^4-alkylation was performed with alkyl halides in the stage of linear intermediates **77**. Furthermore, the hydrophilicity of the polyamide support required an optimization of the

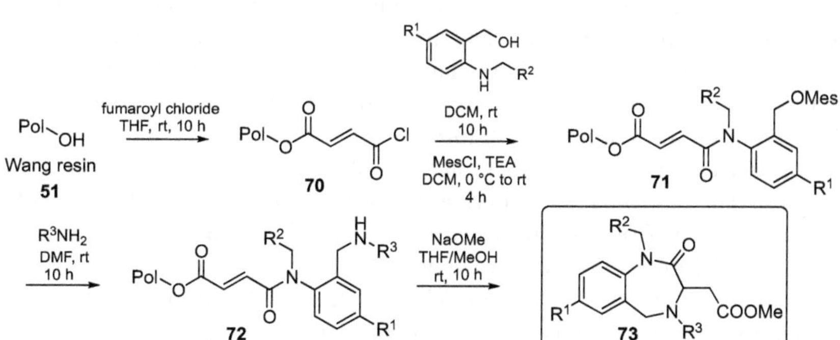

R^1: H, Cl, MeO, Br
R^2: Ph, 4-Me-Ph, 4-MeO-Ph, Me, *i*-Pr, *t*-Bu
R^3: Ph, 4-Cl-Ph, 2-thienyl, furan-2-yl, 4-F-Ph
Yield: above 72%

Scheme 13 Synthesis of 1,4-benzodiazepines by Bhalay et al.

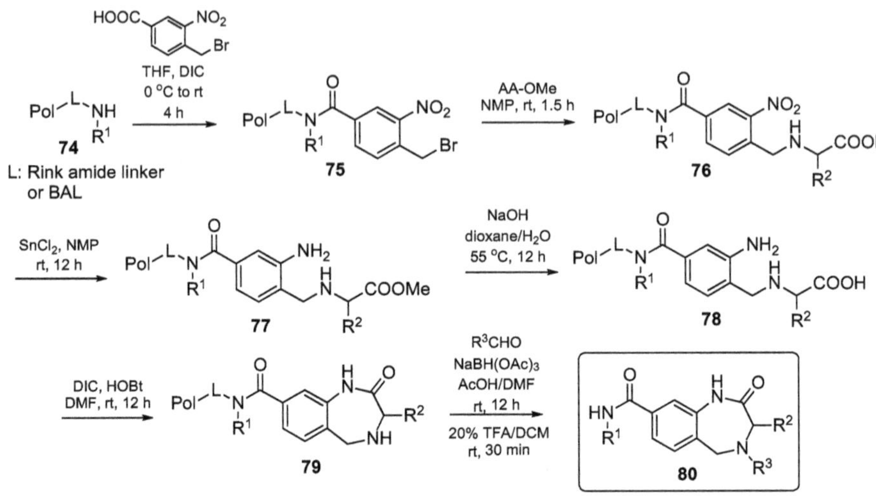

R^1: H, PhEt, pyrrolidin-2-one-*N*-propyl
R^2: *i*-Bu, *i*-Pr, 4-HO-Bn, indol-3-CH_2-
R^3: 4-HO-Bn, Bn, Bu, *c*-Pr-CH_2-, Pyr-3-CH_2-
Yield: 80-95%

Scheme 14 Preparation of benzodiazepines by Lou et al.

reduction step, revealing that the use of $SnCl_2.2H_2O/NH_4OAc$ in a mixture of water and ethanol was the preferred method.

In 2003, Hone published an alternative method based on the combination of solution-phase/solid-phase chemistry [37]. First, the benzodiazepine N^1-Boc-8-carboxylic acid **83** was synthesized in the solution phase according to Lou's procedure and was subsequently attached to the oxime resin **82**. N^4- alkylation with alkyl halides was followed by the cleavage of the Boc-protecting group and N^1 -alkylation with various electrophiles. The aminolysis of resin **86** led to the release of the products from the resin, and the corresponding carboxamides **87** were obtained (Scheme 15). The authors claimed the use of this method for the preparation of a chemical library containing more than 1,300 compounds.

Kim reported the synthesis of β-turn mimetics based on the immobilization of the benzodiazepine building block and its subsequent modification on the solid phase [38]. The benzodiazepine derivative **90** was pre-synthesized in the solution phase in four steps from 2-nitro-5-hydroxybenzaldehyde **88**. After the attachment of **90** to the 4-formyl-3,5-dimethoxyphenoxy (PL-FDMP) resin **89** by reductive amination, the pivaloyl intermediate **91** was unmasked and alkylated with alkyl halides employing LiO^{t}Bu as a base (Scheme 16). The final compounds **93** were released from the resin using TFA, and the crude products were passed through strong anion exchange (SAX) resins to remove TFA. Later, the method was used for the combinatorial synthesis and biological evaluation of the peptide-binding GPCR-targeted library of 162 discrete compounds [39].

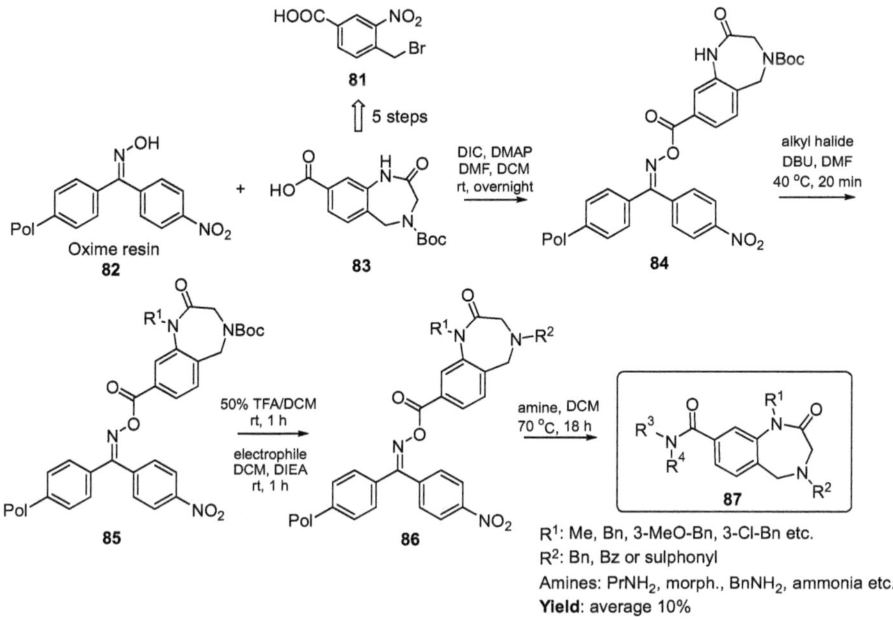

Scheme 15 Combined synthesis of benzodiazepines by Hone et al.

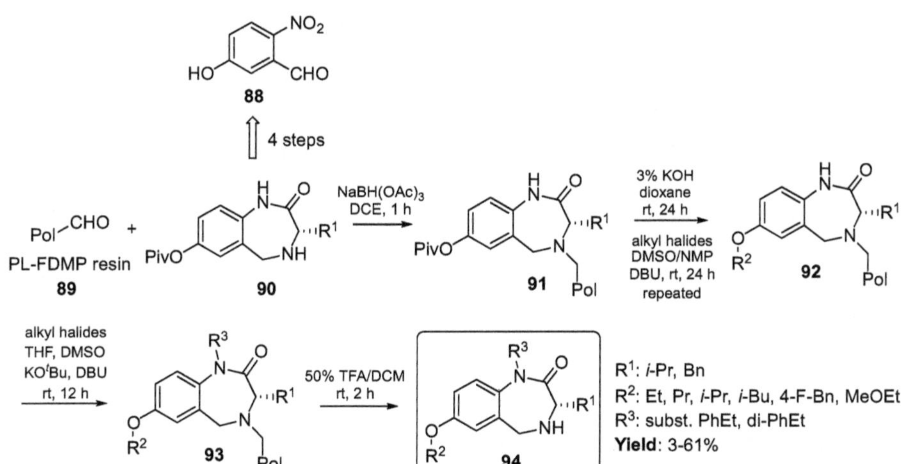

Scheme 16 Synthesis of 7-alkoxy-4-arylalkyl-1,3,4,5-tetrahydro-benzo[*e*][1,4]-diazepin-2-one by Kim et al.

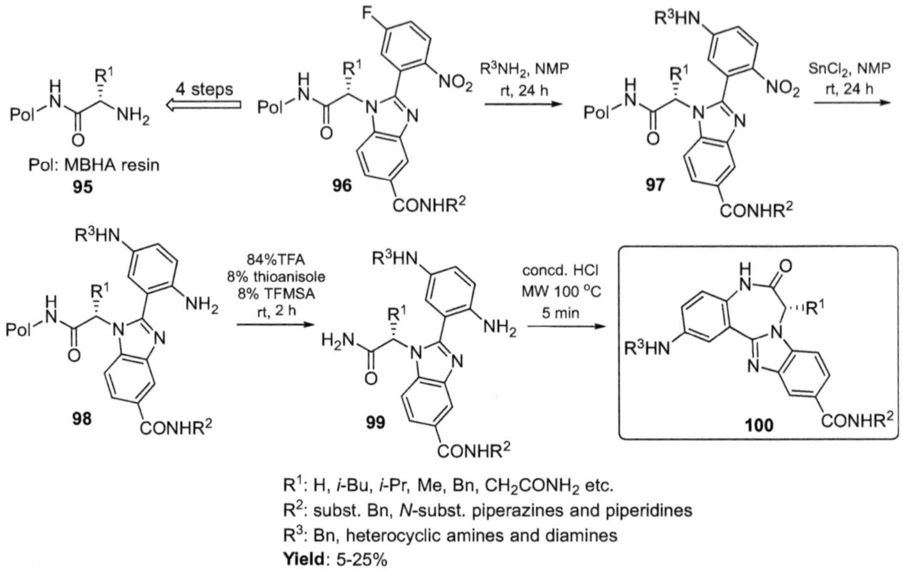

Scheme 17 Benzo[*e*][1,4]diazepin-2-ones fused with benzimidazole scaffolds

The latest contribution devoted to the solid-phase synthesis of tetrahydrobenzo [*e*][1,4]diazepin-2-ones was made by Olsson in 2009 [40]. In a project aimed at novel non-peptidergic MrgX1 and MrgX2 receptor agonists, the preparation of tetracyclic derivatives **100** containing the benzodiazepinone scaffold was developed. Starting from amino acids immobilized on a MBHA resin, polymer-supported benzimidazole intermediates **96** were synthesized in four steps. After replacement of the fluorine substituent with amines followed by the reduction of the nitro group, the linear intermediates **99** were cleaved from resin **98**, and the benzodiazepinones **100** were achieved by short microwave heating in concentrated HCl (Scheme 17). The authors also developed a solution-phase protocol, but the solid-phase alternative was considered superior because the former process required the time-consuming purification of reaction intermediates, which led to significantly reduced yields. The solid-phase strategy was utilized to prepare a chemical library of 500 compounds and to synthesize selected hits in gram quantities.

4.3 *Tetrahydrobenzo[e][1,4]diazepin-3-ones*

The immobilization and solid-phase modification of the 1,4-benzodiazepine-3-one scaffold was reported by Ali with a focus on selective N^4-alkylation [41]. Carboxylic derivative **102** was attached to super acid-sensitive resin (Sasrin) **101** to avoid

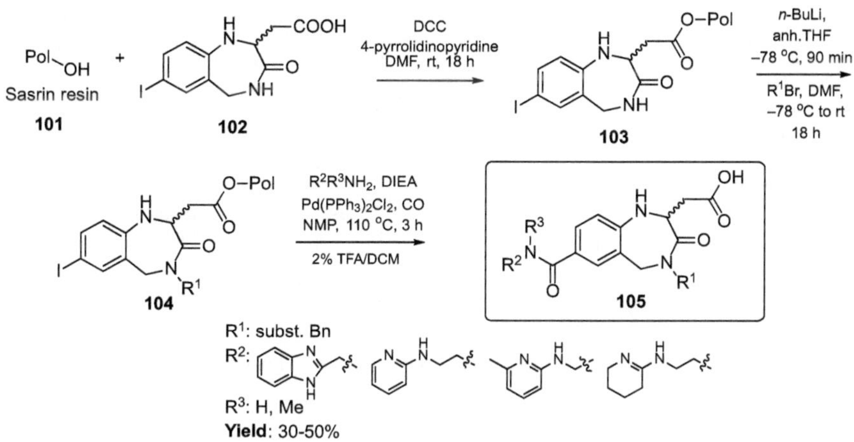

Scheme 18 Regioselective solid-phase N^4-alkylation reported by Ali et al.

benzodiazepine scaffold decomposition due to exposure to strong acids. After selective N^4-benzylation with diverse benzyl bromides, the resin **102** was subjected to Heck coupling, and the final compounds **105** were released from the resin using 1–2% TFA (Scheme 18).

4.4 Pyrrolobenzodiazepines

Pyrrolobenzodiazepines are antibiotic antitumor agents produced by various *Streptomyces* species. These agents bind selectively in the minor groove of DNA via a covalent aminal bond between the C^{11} position of the scaffold and the nucleophilic C^2 amino group of a guanine base, resulting in the observed biological activity [42]. For this reason, pyrrolobenzodiazepines were targeted by medicinal chemists, and numerous contributions devoted to their solid-phase synthesis were reported. The first study was published in 2000 by Thurston, targeting 5-oxo derivatives. The first step involved coupling anthranilic acids to the *p*-nitrophenyl carbonate Wang resin **106** (Scheme 19). The intermediate **107** was coupled with pyrrolidinemethanol to give resin **108**. Following the Dess-Martin oxidation of the hydroxy group, the resin **109** was subjected to cleavage with TFA, and final compounds **110** were isolated [43].

One year later, Kamal reported a method applicable for the preparation of the corresponding 5,11-diones **116** [44]. Instead of anthranilic acids, methyl esters of 2-nitrobenzoic acids were attached to Wang resin **51** via an ether group. After saponification of ester **112**, the polymer-supported carboxylic acids **113** were used for the acylation of L-proline methyl ester (Scheme 20). Reduction of the nitro derivatives **114** led to the formation of the desired scaffold. Prior to acid-mediated

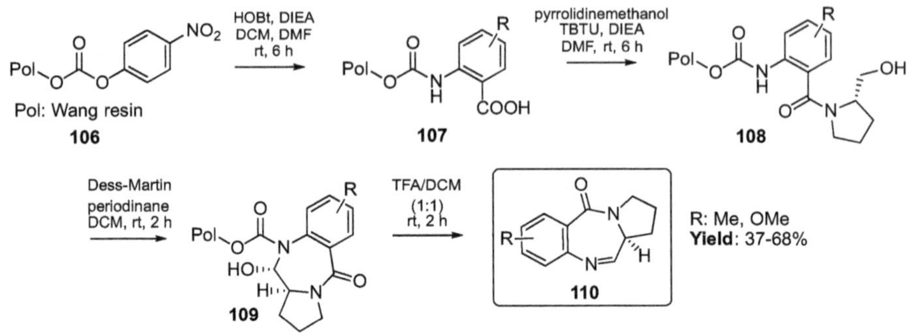

Scheme 19 Preparation of pyrrolobenzodiazepine-5-ones by Thurston

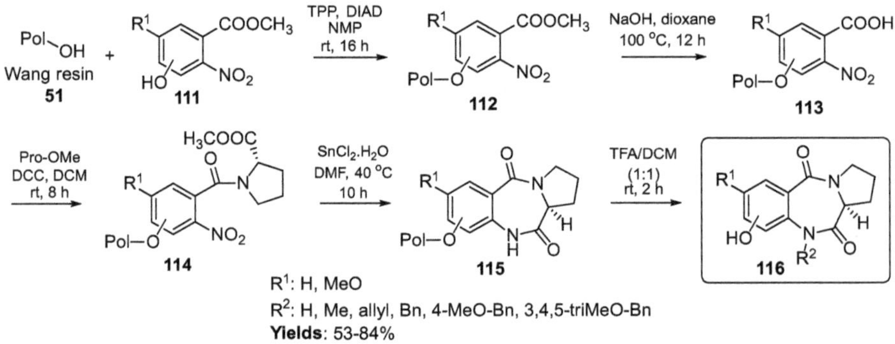

Scheme 20 Preparation of pyrrolobenzodiazepine-5,11-diones by Kamal

cleavage, the second diversity position was created via N^{10}-alkylation with alkyl halides.

Simultaneously, Kamal published an alternative approach consisting of immobilizing the building block **118** via the hydroxy group employing the trichloroacetimidate method (Scheme 21) [45]. Further, reduction with tin(II)chloride dihydrate was replaced with In/NH_4Cl. Despite the advantages of tin reduction, there are examples in the literature wherein substantial quantities of tin by-products remain bound within the resin matrix and are liberated upon acidic cleavage of the desired product [46]. Furthermore, most of the biologically screened cell lines have been shown to be intolerant to tin at these levels. To remove the excess indium, resin **120** was rinsed with water, ethanol, DMF, and DCM. Note that the indium method was also applicable for azido analogues (Scheme 21, X = N_3). In 2009, Santos reported the reductive cyclization of resin **119** with Ni_2B. After microwave irradiation, the products **121** were obtained in good yields, ranging from 64 to 81%, requiring low reaction times [47].

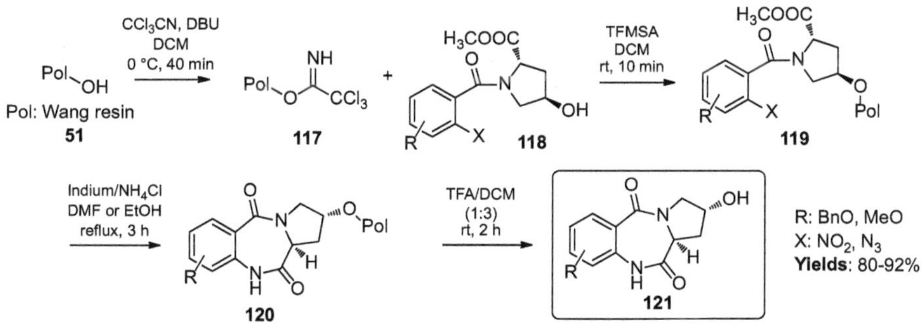

Scheme 21 Preparation of pyrrolobenzodiazepine-5,11-diones via the immobilization of the benzoylproline intermediate

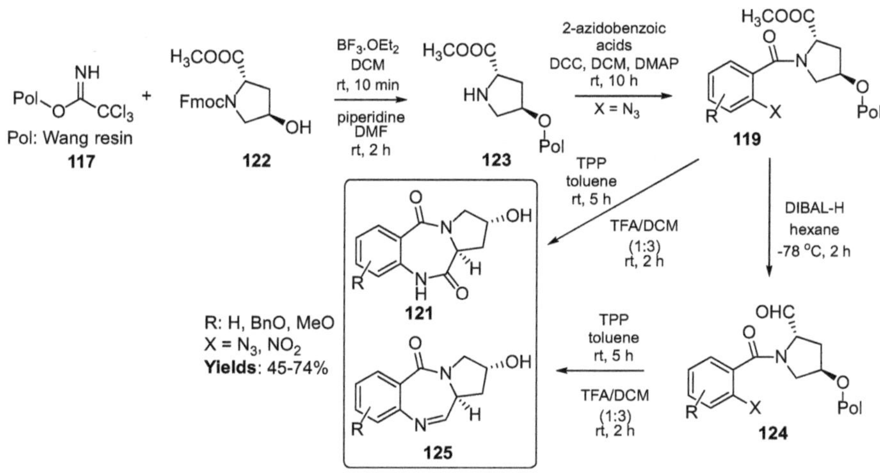

Scheme 22 Preparation of pyrrolobenzodiazepine-5,11-diones and pyrrolobenzodiazepine-5-ones via reduction with TPP

To avoid the necessary solution-phase pre-synthesis of intermediates **118** (Scheme 21), the reaction sequence was further modified employing 4-hydroxy-Fmoc-Pro methyl ester as the building block [48]. After immobilizing 4-hydroxy-Fmoc-Pro methyl ester on the modified Wang resin **117** and after cleavage of the Fmoc-protecting group, resin **123** was acylated with 2-azidobenzoic acid. Reduction with triphenylphosphine (TPP) followed by TFA cleavage afforded the pyrrolobenzodiazepine-5,11-diones **121**. Alternatively, reduction of methyl ester **119** with diisobutylaluminum hydride (DIBAL-H) led to the corresponding aldehydes **124**, which furnished pyrrolobenzodiazepine-5-ones **125** (Scheme 22) upon reduction with TPP. Later, reduction with TPP was successfully replaced with $Al/NiCl_2{\cdot}6H_2O$ and Al/NH_4Cl [49]. As reported, reduction with $Al/NiCl_2$ was faster than that with Al/NH_4Cl, proceeding at room temperature and affording higher yields. Additionally, the reduction method was successfully applied to the corresponding nitro analogues **119** ($X = NO_2$). In 2006, Kamal introduced an

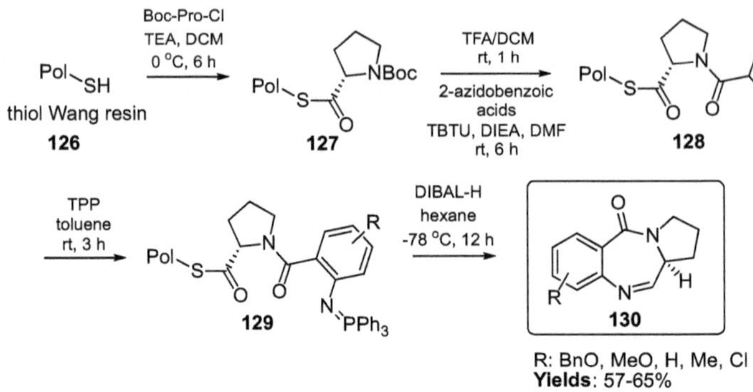

Scheme 23 Preparation of pyrrolobenzodiazepine-5-ones by reductive cleavage

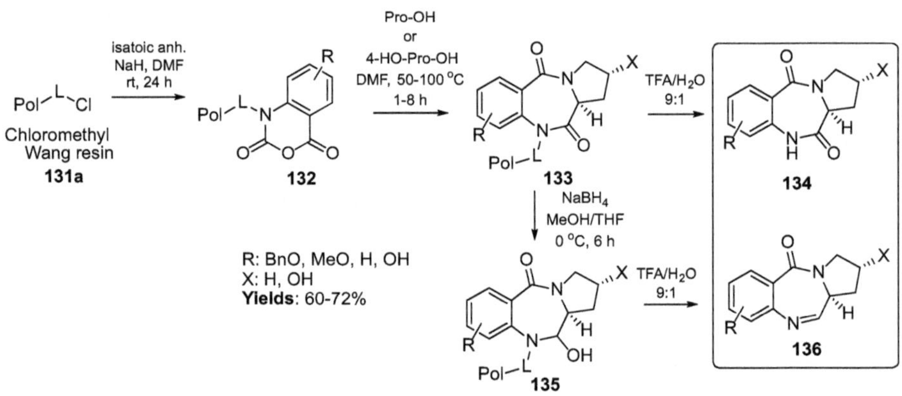

Scheme 24 Preparation of pyrrolobenzodiazepines from polymer-supported isatoic anhydrides

alternative reduction of azido intermediates with boron trifluoride diethyletherate and ethanethiol [50]. The same author reported direct conversion of intermediates **119** to products **121** employing an azido-reductive cyclization approach using $AlCl_3$ in combination with NaI [51].

When thiol Wang resin **126** was used in combination with Boc-Pro-Cl, the proline intermediate was attached to the resin as the thioester **127** [52]. After cleavage of the Boc-protecting group followed by acylation with 2-azidobenzoic acids, the corresponding iminophosphoranes **129** were obtained using TPP. Treatment of resin **129** with DIBAL-H resulted in the reductive cleavage to release the final compounds **130** from the resin, which could be reused for the preparation of intermediate **127** (Scheme 23).

In addition to anthranilic and 2-nitrobenzoic acids, immobilized isatoic anhydrides **132** were also used to construct the pyrrolobenzodiazepine scaffolds **134**. *N*-alkylation of isatoic anhydrides with chloromethyl Wang resin **131a** was followed by reaction with proline, which directly furnished the polymer-supported pyrrolobenzodiazepine-5,11-diones **133** (Scheme 24). Later, the same strategy

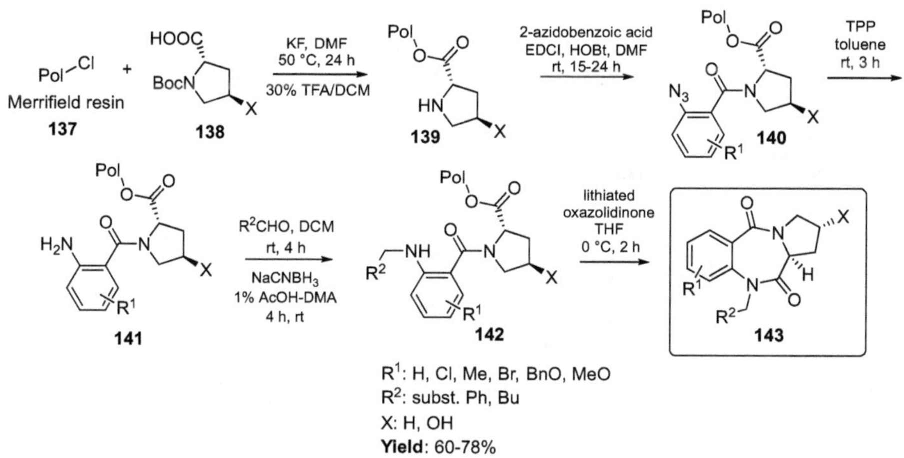

Scheme 25 Preparation of pyrrolobenzodiazepine-5,11-diones via cyclative cleavage with lithiated 5-phenyl-2-oxazolidinone

was applied for the preparation of analogical benzodiazepinones bearing thiophene instead of a benzene ring, with the corresponding thieno-oxazine-2,4(1*H*)-dione as the key building block [53]. Furthermore, to access pyrrolobenzodiazepine-5-ones, resin **133** was subjected to reduction with $NaHB_4$ or $LiBH_4$, followed by the acid-mediated dehydration and cleavage from resin **135** to give derivatives **136**. The isatoic anhydride method was later applied by Waldmann for a solution-phase synthesis of the selected pyrrolobenzodiazepine-5,11-dione, which was immobilized on a chlorotrityl resin and modified to mimic the lipidated C-terminus of the H-Ras protein [54].

The largest chemical library of 210 pyrrolobenzodiazepinediones was reported in 2007 to study antitubercular activity of target compounds [55]. Employing Merrifield resin **137** and Boc-Pro-OH or 4-hydroxy-Boc-Pro-OH **138**, the intermediates **139** were synthesized in a similar manner to the previously described protocols. Reductive alkylation of anilines **141** was followed by treatment of the resin with lithiated 5-phenyl-2-oxazolidinone to afford the desired products **143** (Scheme 25). Although a variety of different bases, such as K_2CO_3, KO^{t}Bu, or NaOMe, also triggered the cyclative cleavage, the oxazolidine method gave the best results.

Similar approaches to those described above were used for the solid-phase synthesis of pyrrolo[2,1-*c*][1,4]benzodiazepine dimers [56] or pyrrolobenzodiazepine-chalcone conjugates [57].

4.5 *Benzo[e][1,4]diazepin-5-ones*

The solid-phase synthesis of benzo[*e*][1,4]diazepin-5-ones was first reported by Park in 2007 [58]. Starting from bromoacetal resin **144**, primary amines were

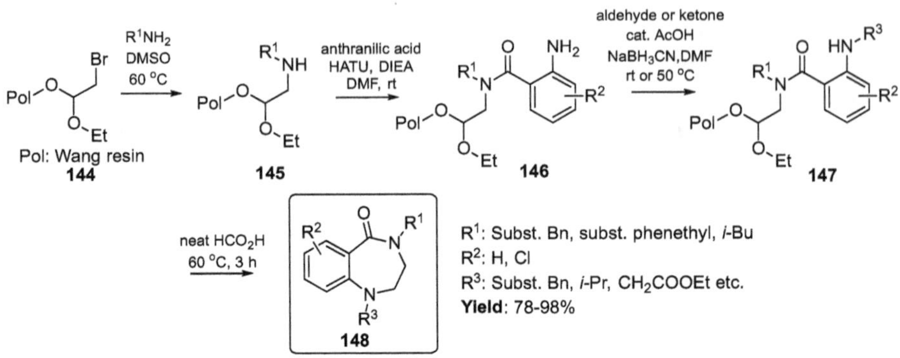

Scheme 26 Preparation of benzo[*e*][1,4]diazepin-5-ones using the Leuckart-Wallach reaction

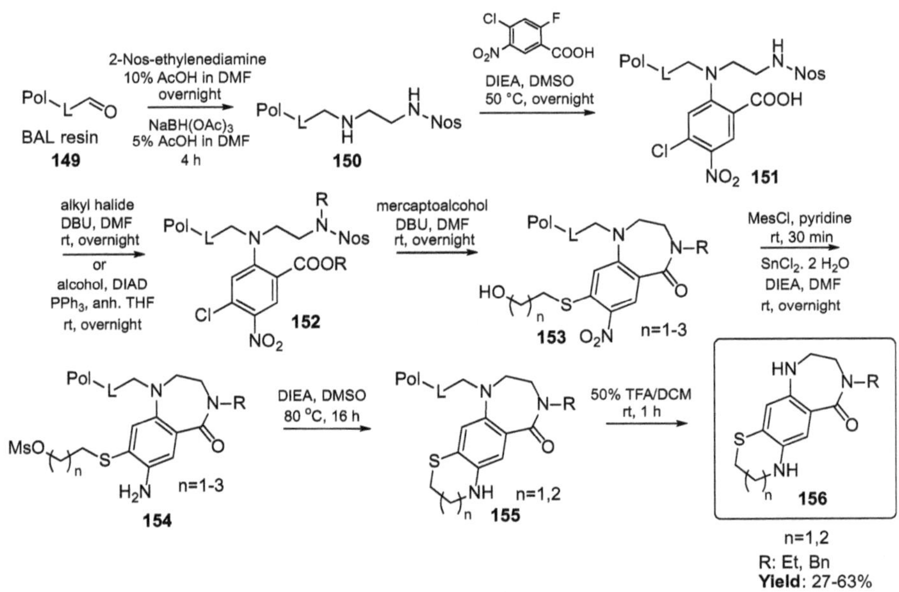

Scheme 27 Preparation of benzo[*e*][1,4]diazepin-5-ones using polymer-supported ethylenediamine

immobilized and acylated with anthranilic acids. After reductive alkylation of resin **146**, the target benzodiazepines **147** were formed by the Leuckart-Wallach reaction (Scheme 26). The robustness and practicality of the synthetic pathway were validated by the successful construction of a 96-member pilot library with excellent overall yields and purities.

In 2012, an alternative method was reported starting from Nos-protected ethylenediamine immobilized by the reductive amination of BAL resin **149** [59]. Intermediate **150** was arylated with 2-fluoro-4-chloro-5-nitrobenzoic acid, followed by the alkylation of resin **151,** which led to parallel esterification of the

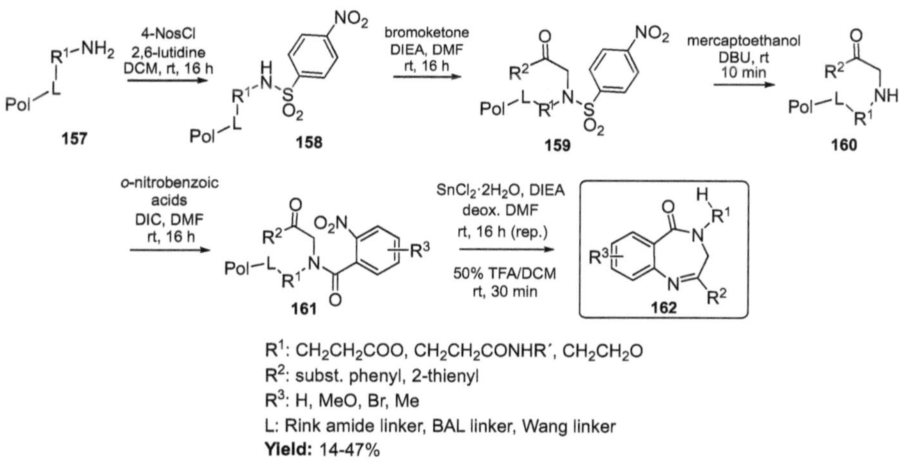

Scheme 28 Preparation of benzo[*e*][1,4]diazepin-5-ones from bromoketones and 2-nitrobenzoic acids

carboxylic group and isolation of resin-bound **152**. A reaction with mercaptoalcohols released the Nos-protecting group, which caused the spontaneous ring closure of benzodiazepine **153** (Scheme 27). Subsequently, the fused heterocycles **156** were prepared via mesylation and reduction of the nitro derivative **153**. The final cyclization of polymer-supported intermediates **154** was successful only for six- and seven-membered rings, whereas the corresponding thiazocines (n = 3) were not detected.

In the same year, the preparation of 3,4-unsaturated benzo[*e*][1,4]diazepin-5-ones **162** was reported [60]. Various polymer-supported amines **157** were reacted with 4-nitrobenezenesulfonyl chloride (4-Nos-Cl) and subjected to alkylation with bromoketones. Cleavage of the 4-Nos derivatives **159** enabled acylation of the resulting aminoketones **160** with 2-nitrobenzoic acids. Note that the acylation step resulted in partial cleavage of the phenacyl group. After reduction of the nitro derivative **161**, the seven-membered scaffold was spontaneously formed on the resin, and the final acid-mediated cleavage afforded the target benzodiazepines **162** (Scheme 28). The use of 2-nitrobenzenesulfonyl chlorides (2-Nos-Cls) resulted in intermediate **159** with the nitro group in the *o*-position, and then, reduction and cyclization afforded the corresponding benzothiadiazepine 1,1-dioxide derivatives [61]. Furthermore, employing 2-Nos-Cls and benzothiadiazepine 1,1-dioxides was recently obtained from polymer-supported α-amino acids using alcohols as the building blocks for the Fukuyma-Mitsunobu N^2-alkylation [62].

The previously mentioned method was further modified to prepare triazolobenzodiazepinones and triazolobenzothiadiazepine 1,1-dioxides [63]. Starting from the intermediates **158**, the alkylation with propargyl bromides was performed. Cleavage of the Nos group followed by the acylation of resin **164** with 2-azidobenzoic acids led to a spontaneous, catalyst-free Huisgen 1,3-dipolar reaction that yielded triazole derivatives **166** (Scheme 29, X = CO). When

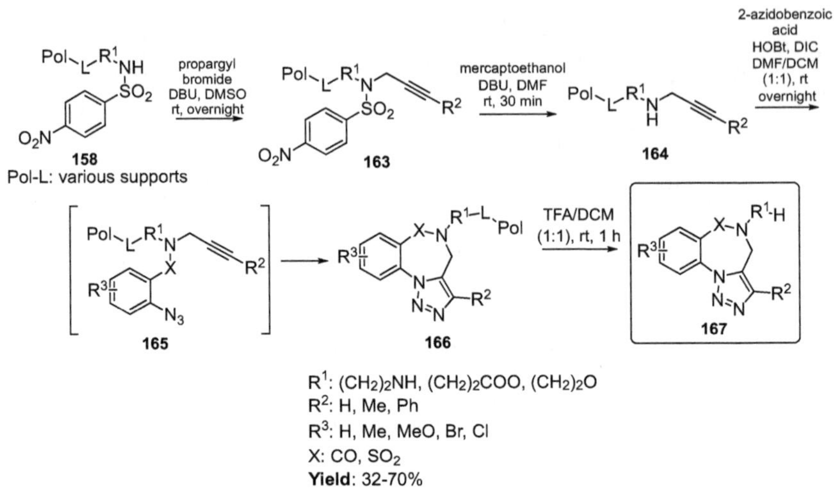

Scheme 29 Solid-phase synthesis of benzotriazolo[1,4]diazepin-6(5*H*)-ones and their sulfonyl analogues

2-azidobenzoic acids were replaced with 2-azidobenzenesulfonyl chloride, the corresponding thiadiazepine1,1-dioxides were obtained (X = SO_2). A similar reaction was described to prepare *N*-unsubstituted triazolobenzodiazepinones with basic alumina as the solid support. However, Cu(phen)(PPh_3)Br was required as the catalyst to trigger triazole formation [64].

4.6 Benzo[e][1,4]diazepine-2,3-diones

Starting from a MBHA resin with immobilized amino acids **168**, benzo[*e*][1,4] diazepine-2,3-diones **172** were synthesized in four steps [65]. First, reductive alkylation with benzaldehydes was performed. The immobilized secondary amines **169** were then treated with methyl chlorooxoacetate to provide the corresponding resin-bound methyl aminooxoacetate **170**. After reduction of the nitro group, benzodiazepine derivatives **171** were spontaneously formed on the resin (Scheme 30). This method was successfully used for the preparation of a chemical library consisting of 200 compounds with yields ranging from 65 to 96%.

4.7 Benzo[e][1,4]diazepine-2,3,5-triones

In 2015, Nefzi reported the preparation of benzo[*e*][1,4]diazepine-2,3,5-triones tethered with a benzimidazole moiety [66]. Similar to the construction of the

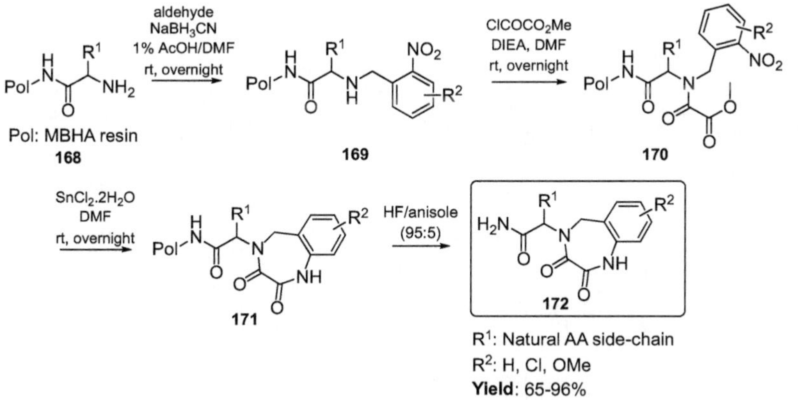

Scheme 30 Solid-phase synthesis of benzo[*e*][1,4]diazepine-2,3-diones

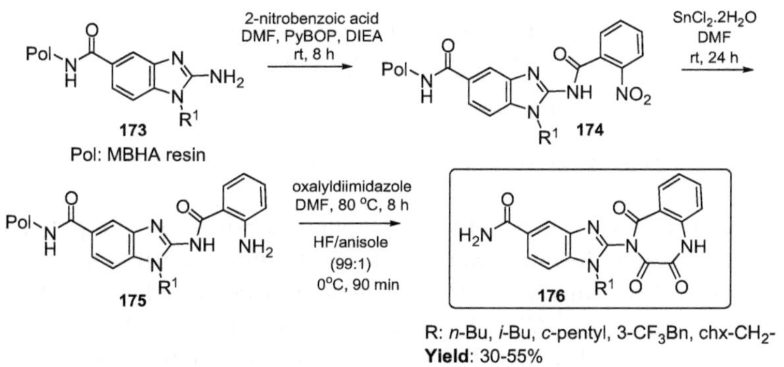

Scheme 31 Solid-phase synthesis of benzo[*e*][1,4]diazepine-2,3,5-triones

benzodiazepinone-2,3-dione scaffold (see Sect. 4.6), an oxalic acid derivative was used to access the seven-membered ring. To prepare linear intermediates **175**, the 2-aminobenzimidazole resin **173** was acylated with 2-nitrobenzoic acid, followed by the reduction of the nitro group. Treatment with oxalyldimidazole furnished the final bisheterocycles **176** that were cleaved from the resin using HF (Scheme 31). Alternatively, intermediate **175** was reacted with 1,1-carbonyldiimidazole to generate the isocyanate derivative, which underwent intramolecular cyclization to furnish the resin-bound quinazoline-2,4-diones.

4.8 Benzo[b][1,4]diazepin-2-ones

The first solid-phase synthesis of benzo[*b*][1,4]diazepin-2-ones was reported by Schwarz [67]. ArgoGel-Rink-resin **177** was acylated with 4-fluoro-3-nitro-benzoic

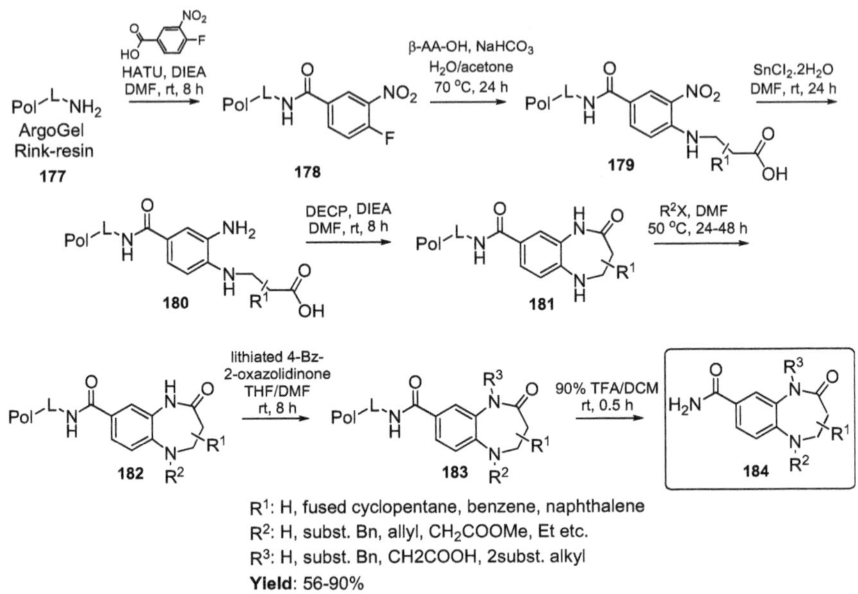

Scheme 32 Solid-phase synthesis of benzo[*b*][1,4]diazepin-2-ones by Schwarz

acid, and resin **178** was subjected to nucleophilic substitution with β-amino acids. After the reduction of the nitro group, the benzodiazepine scaffold was obtained by the cyclization of intermediate **180** using diethyl cyanophosphonate (DECP). The resin was subsequently N^1/N^4 dialkylated to give the final resin-bound product **183** (Scheme 32). The alkyl halides for N^4 encompassed over forty benzyl bromides, allyl bromides, and bromoacetic acid esters, as well as methyl and ethyl iodide. However, other alkyl iodides, along with benzyl chlorides and α-bromo-acetophenones, did not exhibit satisfactory results. The N^1-alkylation was accomplished using alkyl halides with lithiated 4-benzyl-2-oxazolidinone as a base. From the spectroscopic data of the final products, no evidence was found for C- and/or O-alkylation. In 1999, the same approach was reported by Lee, with an application of the Rink amide resin and β-amino acid methyl esters [68]. Note that the authors did not manage to cyclize the benzodiazepine scaffold via the intramolecular aminolysis of the ester functionality; therefore, the subsequent hydrolysis and HOBt cyclization had to be performed. However, the on-resin cyclization with ethyl ester was successfully performed by heating in TFA when soluble PEG resin was used as the polymer support [69].

In 2000, Herpin reported the preparation of a large chemical library (10,000 compounds) based on the immobilization of the (benzo[*b*][1,4]diazepin-3-yl) isoindoline-1,3-dione building block via its N^1 position [70]. Alkylation of the N^4 position was followed by the cleavage of the phthalimide **187** and acylation of the primary amino group to give the final trisubstituted products **189** (Scheme 33).

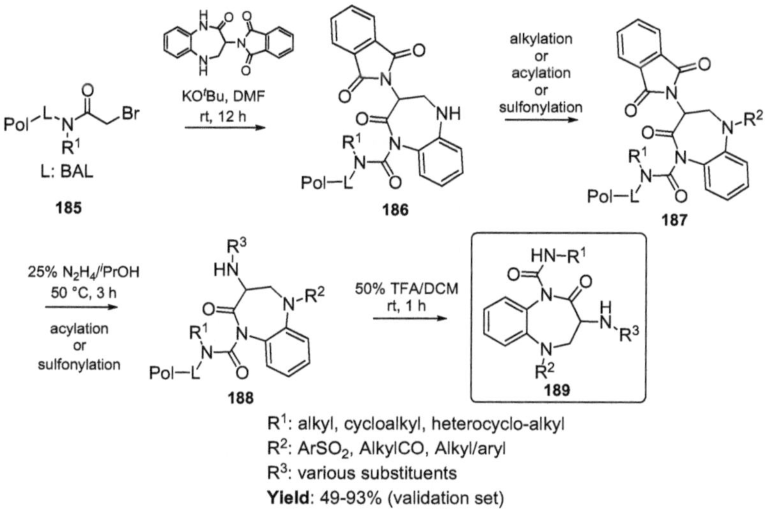

Scheme 33 Immobilization and on-resin modification of benzo[*b*][1,4]diazepin-2-one scaffold by Herpin et al.

Finally, the preparation of benzo[*b*][1,4]diazepin-2-ones was reported by Kidway from *o*-phenylenediamines and α,β-unsaturated carboxylic acids employing acidic alumina as the solid support [71].

4.9 *Benzo[b][1,4]diazepines*

The use of polystyrene/1% divinylbenzene sodium sulfinate **190** enabled the preparation of different heterocycles, including 2,3-dihydro-1*H*-benzo[*b*][1,4] diazepines [72]. Alkylation followed by the reaction of resin **191** with epoxides led to secondary alcohols **192**, which upon oxidation with Jones reagent, gave the key intermediates **193**. The reaction of these intermediates with *o*-phenylenediamine in toluene in the presence of TEA yielded the product **194** in 10–38% overall yield (Scheme 34). Although the reaction using KOH/ethanol, KOH/DMA, or KOH/DMF instead of TEA/toluene proceeded more rapidly, it gave poorer yields. Four compounds were prepared in this manner.

Recently, the reaction of *o*-phenylenediamine and α-oxo ketene dithioacetals with basic alumina as the solid support to prepare benzo[*b*][1,4]diazepines was reported by Bhagat [73].

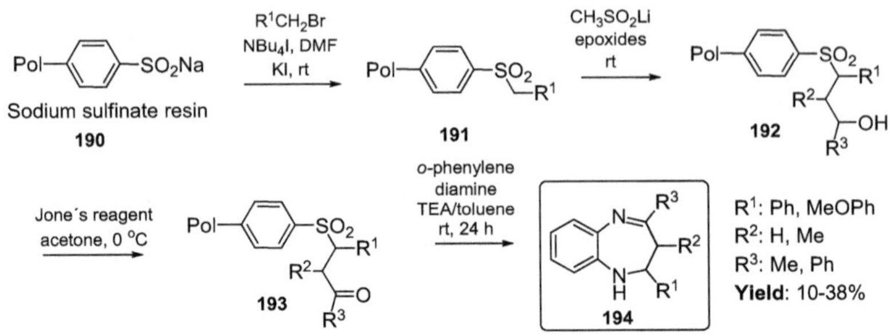

Scheme 34 Use of a sodium sulfonate resin for the preparation of 2,3-dihydro-1*H*-benzo[*b*][1,4] diazepines

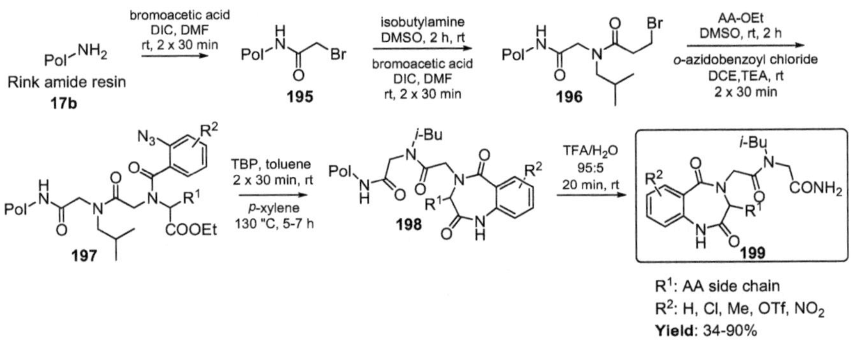

Scheme 35 The first solid-phase synthesis of benzo[*e*][1,4]diazepine-2,5-diones

4.10 Benzo[e][1,4]diazepine-2,5-diones

Along with the 1,3-dihydro-2*H*-benzo[*e*][1,4]diazepin-2-ones discussed in Sect. 4.1, benzo[*e*][1,4]diazepine-2,5-diones have been the most frequently reported [1,4]diazepines in solid-phase synthesis. The first paper from this field of research was published in 1995 by Goff et al. [74] Starting with the Rink amide resin **17b** acylated with bromoacetic acid **195**, the intermediate **196** was obtained via immobilization of isobutylamine. Repeating the acylation step with bromoacetic acid followed by reacting with amino acid esters afforded a resin that was acylated with 2-azidobenzoyl chlorides. Treatment of resin-bound **197** with tributylphosphine (TBP) gave the iminophosphorane, which was heated to yield the benzodiazepine intermediate **198** that is cleavable by TFA (Scheme 35). This method was successfully tested for a wide range of amino acids with only minor limitations, which are the use of L-asparagine *tert*-butyl ester and trityl-protected L-histidine methyl ester and sterically hindered L-valine methyl ester. In 2008, Goff's strategy was further modified by Subra for the online synthesis of a pseudopeptide library incorporating benzodiazepinone as a mimic with a subsequent biological evaluation of the chemical library on melanocortin-1 (MC1) receptors [75]. Instead of

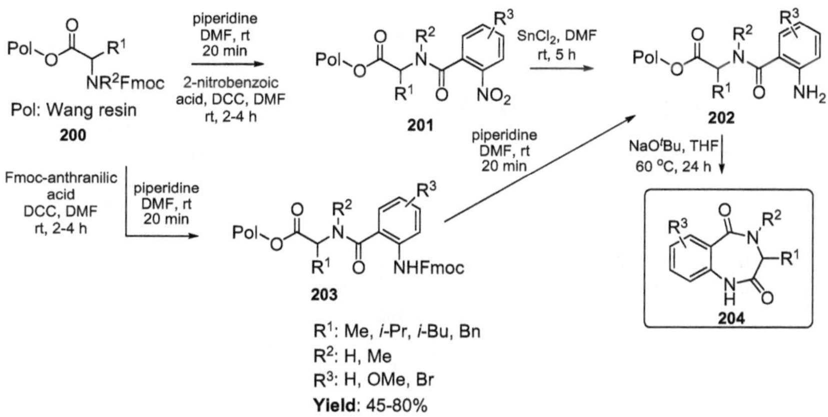

Scheme 36 Solid-phase synthesis of benzo[*e*][1,4]diazepine-2,5-diones by Mayer

2-azidobenzoyl chlorides, 2-nitrobenzoyl chlorides were used, and resin-bound benzodiazepines were obtained after the reduction of the nitro group [76].

One year after Goff's finding, a simple method was reported by Mayer [77]. Fmoc-amino acid-derivatized Wang resin **200** was subjected to cleavage of the Fmoc-protecting group, and the resulting resin was acylated with 2-nitrobenzoic acids, followed by the reduction of the nitro group (Scheme 36). Alternatively, the intermediates **202** were obtained via the acylation with Fmoc-anthranilic acids and deprotection of resin **203** with piperidine. This methodology was later utilized by Waldmann for direct on-bead monitoring of the solid-phase synthesis of benzodiazepines **204** using soft laser desorption time-of-flight mass spectrometry (SLD-TOF MS) without prior cleavage from the resin using a photocleavable linker [78]. A number of cyclization conditions to prepare products **204** was evaluated by Mayer, and optimal results were obtained by heating the resin **202** in tetrahydrofuran sodium-*t*-butoxide. The cyclative release feature of this approach resulted in a significant enhancement of the purity of the final compounds. In 2003, Mayer's method was applied by Migihashi to prepare a chemical library of 400 benzodiazepines employing an ACT-496 automatic synthesizer and IRORI radio-frequency-encoded split-mix synthesis technology [79]. Later, Fmoc-anthranilic acids were replaced with Boc-anthranilic acids, and a library of benzodiazepines was synthesized on a Wang resin [80] or on a Kaiser oxime resin [81]. Compared with Mayer's approach, the final cleavage with TFA/DCM resulted in a significantly easier work-up of the crude products based on simple evaporation of the cleavage cocktail.

An interesting methodology to synthesize benzodiazepines **204** using solid-phase synthesis from anthranilic acids is based on a multicomponent Ugi reaction. Hulme was the first to report this strategy using a safety-catch linker [82]. The isonitrile resin **205** was prepared from a Wang resin in four steps and subjected to a one-pot reaction with aldehydes, amines, and Boc-anthranilic acid to give the resin-bound intermediate **206**. Boc derivatization promoted facile cleavage from the resin **207** with methoxide, giving the corresponding methyl esters **208**. Finally, cleavage

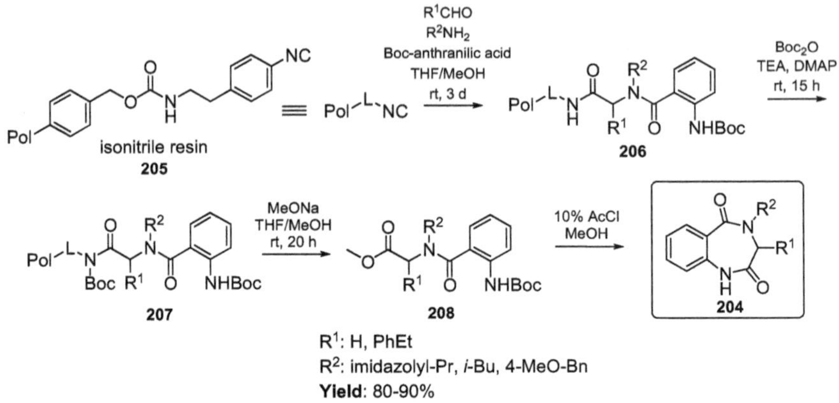

Scheme 37 Ugi reaction and an isonitrile resin for the preparation of benzo[*e*][1,4]diazepine-2,5-diones

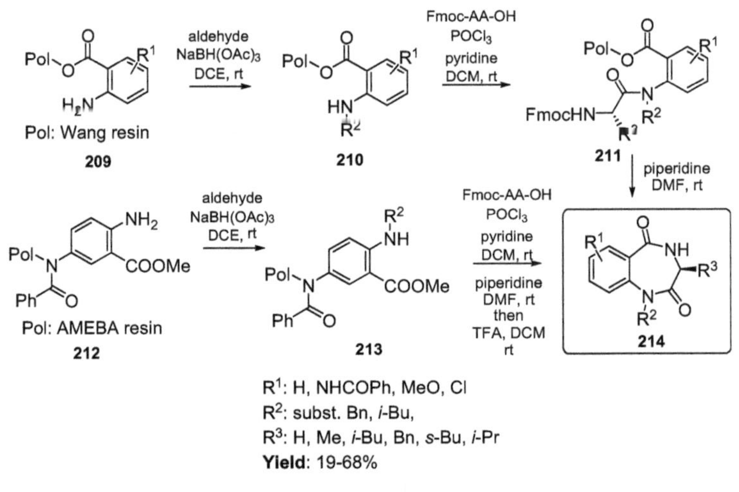

Scheme 38 Solid-phase synthesis of benzo[*e*][1,4]diazepine-2,5-diones from immobilized anthranilates

of the Boc-protecting group from anthranilate **208** was followed by cyclization, yielding benzodiazepines **204** (Scheme 37). Later, the application of the Ugi reaction to obtain benzodiazepines on a solid phase from Fmoc-anthranilic acids [83] or *N*-substituted anthranilic acids [84] was also reported.

A reversed immobilization scenario was reported by Jeon et al. [85]. Instead of amino acids, anthranilic acids were attached to a Wang resin. Two approaches were developed: (a) immobilization via an anthranilic acid carboxylate (resin **209**) and (b) immobilization via an additional functional group of the aromatic moiety (resin **212**). In the first case, polymer-supported anthranilic acids were reductively alkylated to **210**, and upon the acylation with Fmoc-amino acids, the cleavage of

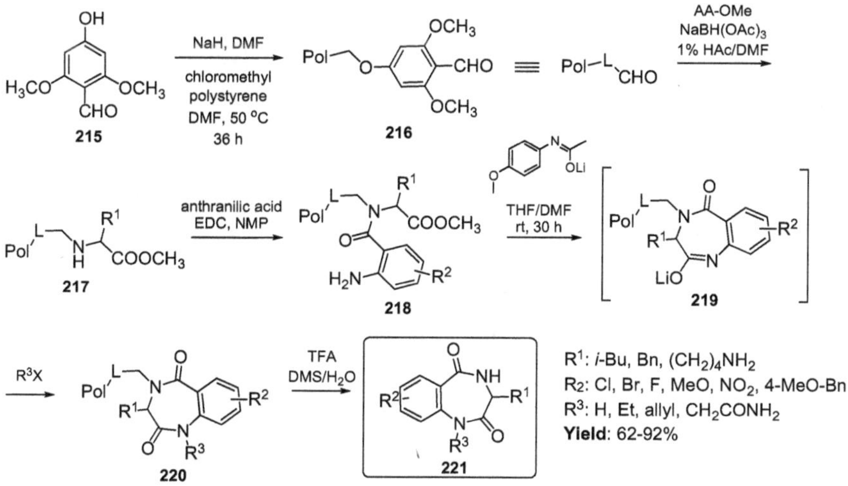

Scheme 39 Solid-phase synthesis of benzo[*e*][1,4]diazepine-2,5-diones by Ellman et al.

the Fmoc-protecting group led to cyclative cleavage, releasing the final compounds **214** from the resin **211**. In the latter case, the resin-bound anthranilic acid methyl ester **212** was reductively alkylated, and the intermediate **213** was coupled with Fmoc-amino acids. Cleavage of the Fmoc-protecting group led to on-resin cyclization of the benzodiazepine product, which was cleaved from the resin using TFA (Scheme 38). Prior to the cleavage, the intermediates were eventually N^4-derivatized with alkyl halides.

A different strategy was described by Ellman [86]. Starting from a chloromethyl resin equipped with benzaldehyde linker **216**, various amino acid methyl esters were immobilized via the N-terminus using reductive alkylation (Scheme 39). The acylation of intermediate **217** with anthranilic acids required considerable optimization. For example, even highly activated reagents, such as HATU, gave poor conversion. Carbodiimides were the only coupling agents found to efficiently cause this transformation. Furthermore, good yields of acylated material were obtained only when the carbodiimides were employed in conjunction with the hydrochloride salt of a tertiary amine. *N*-(3-dimethylaminopropyl)-*N′*-ethylcarbodiimide hydrochloride (EDC) was shown to be the most convenient activating agent due to the presence of the tertiary amine hydrochloride in the carbodiimide structure. The reaction of intermediate **218** with the lithium salt of *p*-methoxyacetanilide followed by the addition of an appropriate alkylating agent provided a fully derivatized, resin-bound benzodiazepine **220**. A similar approach was later used to synthesize 7-acylamino benzo[*e*][1,4]diazepine-2,5-diones [87]. In 2008, Ellman's method was applied for the preparation and screening of a benzodiazepine library to identify novel melanocortin receptor agonists with nanomolar potencies [88].

In 2003, Rivero reported the solid-phase synthesis of benzo[*e*][1,4]diazepine-2,5-diones by the intramolecular cyclization of bromoacetyl intermediate **224**

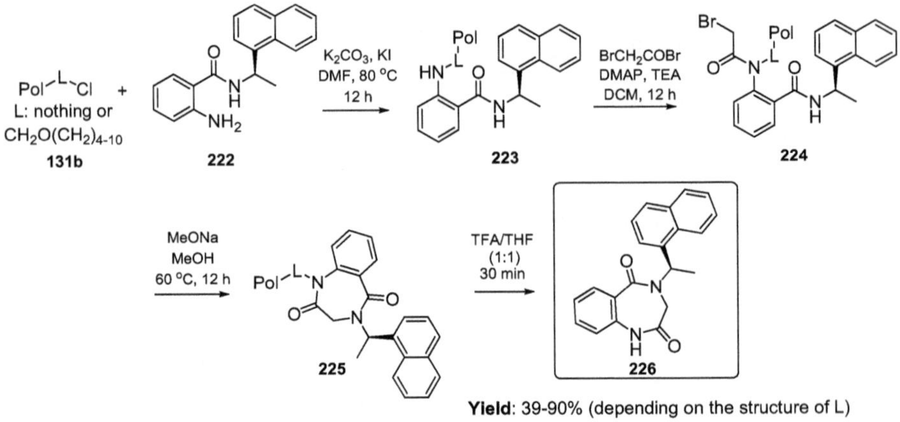

Scheme 40 Solid-phase synthesis of benzo[*e*][1,4]diazepine-2,5-diones by bromoacetyl resin cyclization

[89]. The resin **223** was prepared from anthranilamide building block **222** immobilized on a chloromethyl resin followed by acylation with bromoacetic acid. The ring formation was performed using four different methods: cesium or sodium carbonate in DMF, sodium methoxide in refluxing methanol, and tetramethylguanidine in NMP (Scheme 40). In contrast to low soluble carbonates, sodium methoxide, which is soluble in methanol, considerably increased the yields of ring formation. However, concurrent hydrolysis of benzamide resin **224** was also observed (30%).

4.11 1,4-Diazepane-2,5-diones

Synthetic strategies for the solid-phase synthesis of diazepane-1,4-diones are similar to those reported for analogous benzodiazepines. Polymer-supported amino acids are typically used as the starting material. One of the first studies in this field was published in the late 1990s by Houghten [90, 91]. Starting from immobilized aspartic acid **227**, the amino group was reductively alkylated, followed by the acylation of intermediate **228** with Fmoc-amino acids (Scheme 41). No racemization was observed when the corresponding imine was reduced immediately upon formation. The subsequent cleavage of the Fmoc-protecting group was followed by reductive alkylation and furnished resin-bound linear intermediate **230**; upon cleavage of the Boc-group, **230** was subjected to cyclization with HATU. Good yields were obtained with Phe and Met(O), whereas low yields were obtained with hindered amino acids, such as Val. Dimerization of the unreacted *N*-substituted aspartic acid (i.e., compound **228** without the *t*-Bu) yielding the 1,5-diazocane scaffold was also observed (5–15%). In 2011, Messeguer reported the synthesis

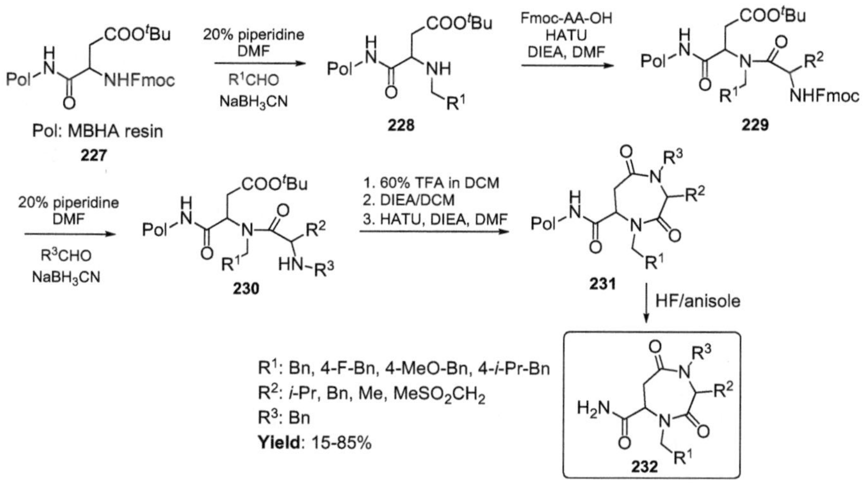

Scheme 41 Solid-phase synthesis of 1,4-diazepane-2,5-diones by Houghten et al.

of intermediates **229** (with allyl instead of Fmoc or Boc) from resin-bound amines by reaction with allyl maleate, followed by an aza-Michael reaction [92]. In this case, the diazepine scaffold was obtained after hydrolysis of allyl ester and on-resin cyclization with benzotriazol-1-yl-oxytripyrrolidinophosphonium hexafluorophosphate (PyBOP).

Simultaneous with the report by Houghten, Krchňák reported an alternative strategy employing Fmoc-Asp(OAll)OH as the acylating species to obtain intermediates **234** from polymer-supported amino alcohols or symmetrical diamines **233** [93]. After reductive alkylation, the R^4 diversity position was introduced via the reaction of resin **235** with bromoacetic acid followed by nucleophilic displacement with primary amines. To remove the protecting group, the allyl ester **236** was treated with $Pd(PPh_3)_4$ or sodium hydroxide, and the cyclization to benzodiazepines **237** was achieved by mild activation of the carboxylate with diphenylphosphoryl azide (Scheme 42). Using this method, a chemical library was synthesized employing the split/mix technique with 8 secondary diamines and amino alcohols, 17 aldehydes, and 20 primary amines, providing 2,720 compounds.

The preparation of 1,4-diazepane-2,5-diones based on a combination of α- and β-amino acids was reported by Amblard [94]. The benzaldehyde resin **239** was reductively aminated with β-alanine benzyl ester, followed by acylation with Fmoc-Phe-OH. After cleavage of the Fmoc-protective group, the resin-bound intermediate **241** was subjected to base-induced cyclization (Scheme 43). The procedure was successfully transferred to a secondary amine, forming derivatized (SAF) Synphase™ crowns. A reversed immobilization strategy was recently reported for the preparation of 1,4-diazepane-2,5-diones **243** fused with a cyclohexane scaffold [95]. The α-amino acids were initially immobilized on a Wang resin via a carboxylate group, and the desired scaffold was formed by cyclative cleavage upon acylation with β-amino acids.

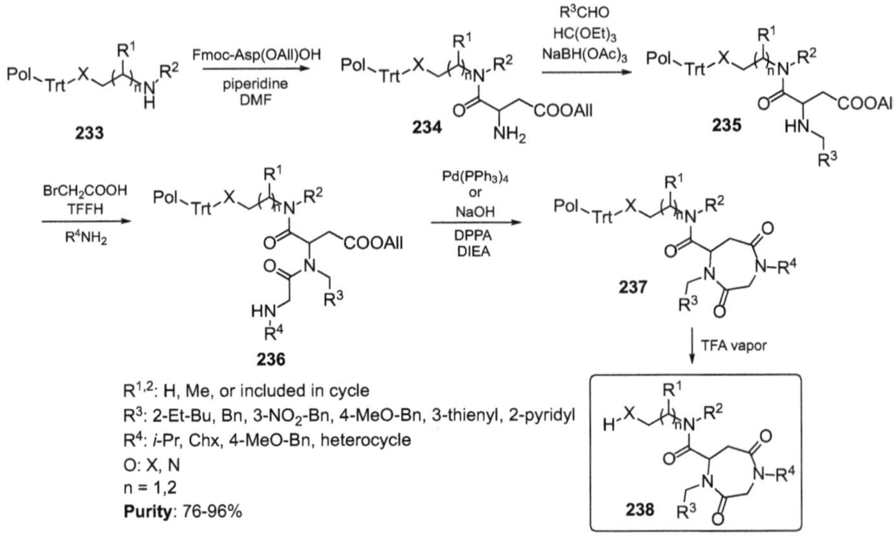

Scheme 42 Solid-phase synthesis of 1,4-diazepane-2,5-diones by Krchňák

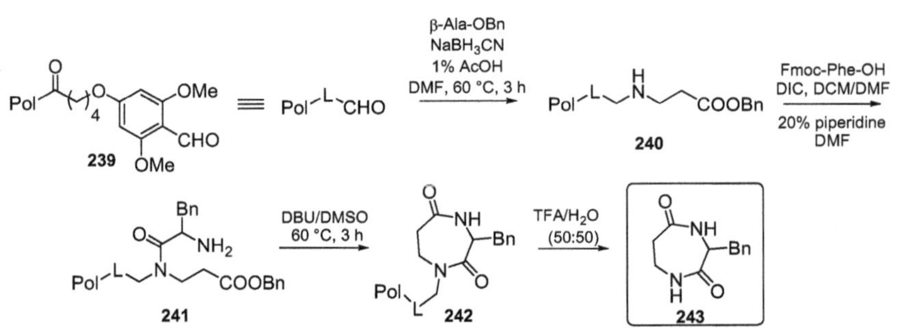

Scheme 43 Solid-phase synthesis of 1,4-diazepane-2,5-diones from α- and β-amino acids

Similarly, it was found that using aspartic acid β-benzyl ester in Fmoc-based solid-phase peptide synthesis not only risks the formation of the aspartimide peptide but also results in its further transformation into a 1,4-diazepane-2,5-dione-peptide derivative [96].

Ring closure of the dipeptide via an intramolecular Mitsunobu alkylation was reported employing hydroxylamine linked to a PS-DVB 2-chlorotrityl resin **244** [97]. Acylation with Fmoc-Phe-OH was followed by the formation of resin-bound dipeptide **246**, and the Mitsunobu reaction yielded the polymer-supported diazepine **247** (Scheme 44). Cyclization was performed with DMF as a solvent in the presence of DIAD and PPh_3 in a sealed tube under microwave irradiation. Three cycles were required to complete the reaction. The reaction pathway was also verified for Fmoc-Glu(OAll)OH as the starting amino acid.

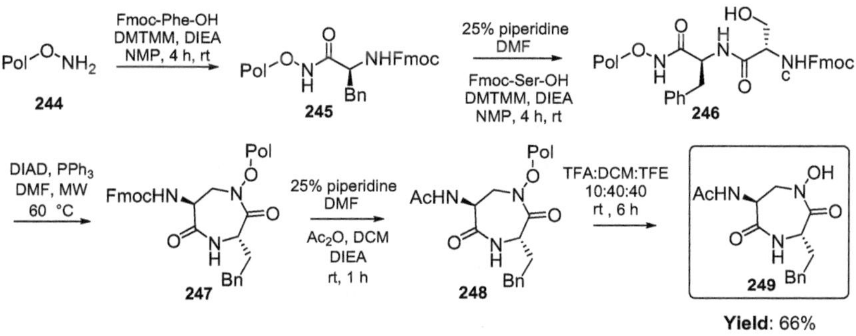

Scheme 44 Solid-phase synthesis of 1,4-diazepane-2,5-diones via the Mitsunobu alkylation

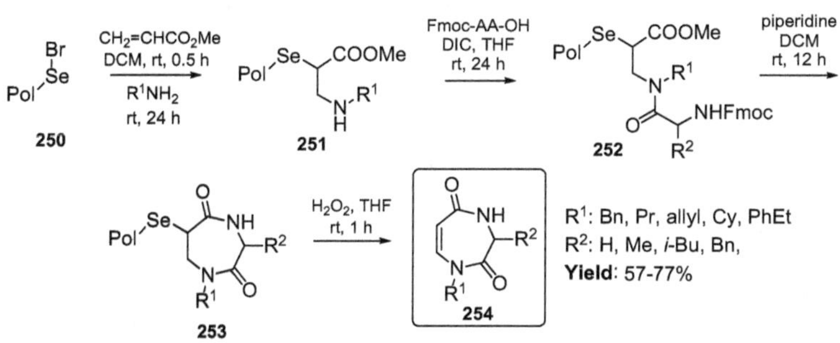

Scheme 45 Solid-phase synthesis of 1,4-diazepine-2,5-diones with a selenyl bromide resin

4.12 1,4-Diazepine-2,5-diones

The solid-phase synthesis of unsaturated analogues **254** of the abovementioned 1,4-diazepane-2,5-diones was reported by Huang who employed resin-bound 3-amino-2-seleno ester **251** [98]. Polystyrene-supported selenyl bromide **250** was successively reacted with methyl acrylate and primary amine in one pot to yield the desired resins. C^2-immobilized β-alanine methyl ester **251** was acylated with Fmoc-amino acids, and after deprotection occurred, spontaneous cyclization gave the resin-bound 1,4-diazepane-2,5-diones **253**. Oxidation and *syn*-elimination led to the cleavage of resin **253** and afforded the final 1,4-diazepines **254** (Scheme 45). When propargylamine was used as the R^1NH_2 building block, the same methodology was applied to synthesize 1,4-diazepine-indole/benzofuran bisheterocycles [99]. Similarly, the use of mono-Boc-*o*-phenylenediamine afforded fused diazepino[1,2-*a*] benzimidazoles [100].

4.13 1,4-Diazepanes

The intramolecular Fukuyama-Mitsunobu alkylation was applied for the preparation of 1,4-diazepanes **258** [101, 102]. The aziridine resin **255** was treated with different aminopropanols, and the resin-bound intermediates **256** were cyclized with triethylphosphine (TEP) and diethyl azodicarboxylate (DEAD) (Scheme 46).

The 1,4-diazepane scaffold was also obtained by the cyclization of polymer-supported diamines **259** prepared from aminomethylpiperidine and a trityl resin in six steps [103]. After the cyclization was performed with propane-1,3-ditriflate, the diazepane **260** was cleaved from the resin and converted to the *N*-pivaloyl product **261** (Scheme 47).

An interesting strategy for the preparation of diimidazodiazepines via solid-phase synthesis was recently reported by Nefzi [104]. The resin-bound tripeptidic intermediate **262** was reduced to the polyamine **263**. After **263** was cleaved from the polymer support, the cyclization with diethyl malonoimidate dihydrochloride afforded diimidazodiazepine compounds **265** (Scheme 48).

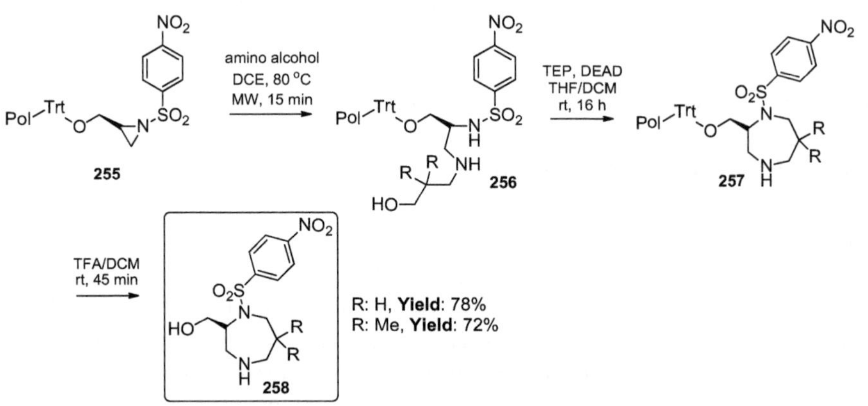

Scheme 46 Solid-phase synthesis of 1,4-diazepanes via the Mitsunobu alkylation

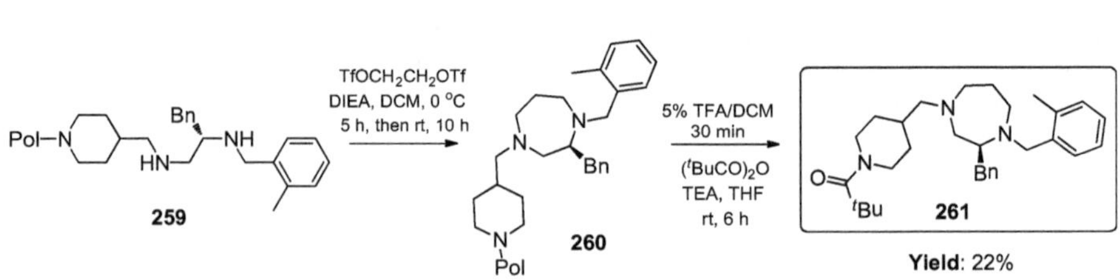

Scheme 47 Solid-phase synthesis of 1,4-diazepanes via cyclization with propane-1,3-ditriflate

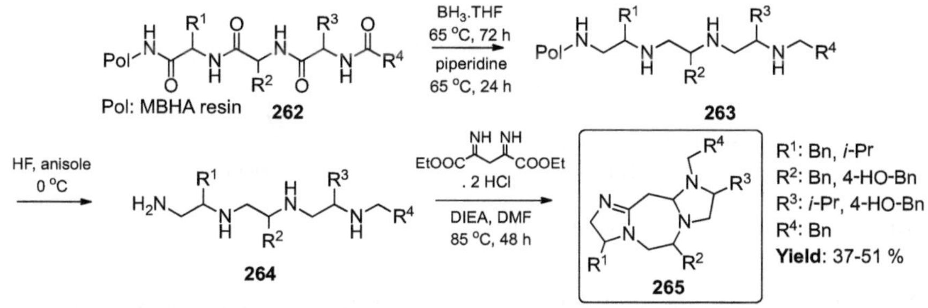

Scheme 48 Synthesis of diimidazodiazepines via solid-phase synthesis

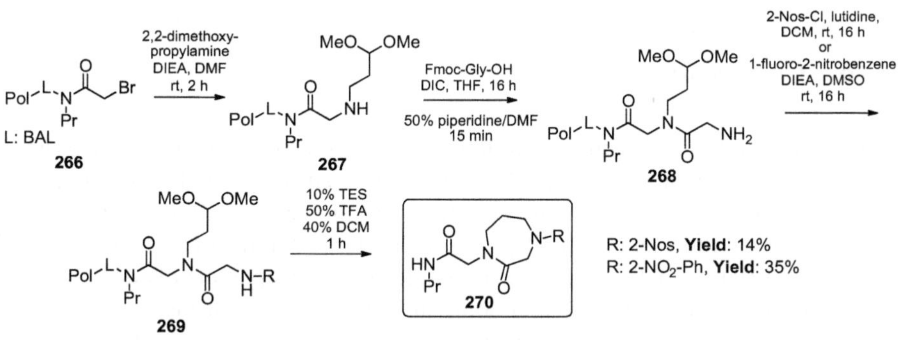

Scheme 49 Synthesis of 1,4-diazepan-2-ones

4.14 1,4-Diazepan-2-ones

Starting from polystyrene resin equipped with a BAL linker, propylamine was immobilized via reductive alkylation followed by acylation with bromoacetic acid. The intermediate **266** was reacted with 2,2-dimethoxypropylamine (Scheme 49). Acylation with glycine afforded the resin **268**, which upon modification of the amino group, was subjected to cleavage with TFA in the presence of triethylsilane (TES). The diazepanones **270** were obtained in limited yield [105]. When the use of TES was omitted, the cleavage yielded the corresponding unsaturated diazepines [106].

4.15 1,4-Diazepan-5-ones

A similar approach to that used in the previous case was applied for the preparation of diazepanones with different positioning of the carbonyl group [105]. The intermediate **271** was synthesized according to Scheme 49 from the Rink amide resin,

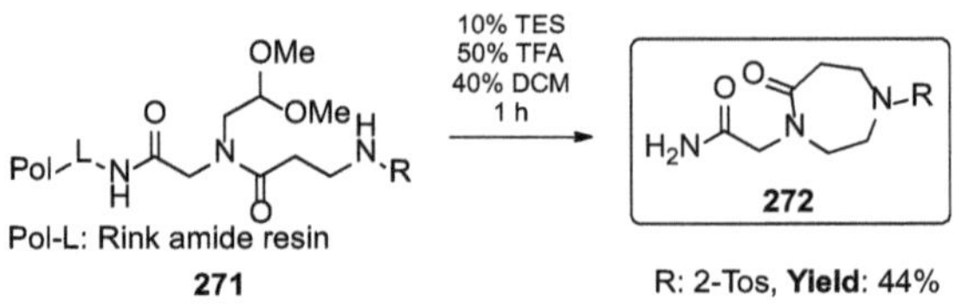

Scheme 50 Synthesis of 1,4-diazepan-5-ones

2,2-dimethoxethylamine and Fmoc-γ-aminobutyric acid. The cleavage afforded 1,4-diazepan-5-one **272** (Scheme 50) or the corresponding 1,4-diazepin-5-ones in the absence of the reducing agent [106].

4.16 Conclusion

The use of solid-phase synthesis for the preparation of pharmacologically promising heterocycles is a challenging field that offers an almost unlimited source of chemical reactions to produce novel, interesting molecules. Although diazepanes and diazepines represent a small group in the large family of heterocycles, the number of different methodologies summarized in this chapter clearly demonstrates the robustness and applicability of solid support chemistry in the area of seven-membered scaffolds. This field has not been fully exploited, and further strategies will be presented to extend the current knowledge.

Acknowledgments The authors are grateful to project CZ.1.07/2.3.00/20.0009 from the European Social Fund and the National Program of Sustainability (project LO1304).

References

1. Charan RD, Schlingmann G, Janso J, Bernan V, Feng X, Carter GTJ (2004) Nat Prod 67 (8):1431–1433
2. Huang RM, Ma W, Dong JD, Zhou XF, Xu T, Lee KJ, Yang X, Xu SH, Liu Y (2010) Molecules 15(2):871–877
3. Hadjipavlou-Litina D, Hansch C (1994) Chem Rev 94(6):1483–1505
4. Butcher JW, Liverton NJ, Claremon DA, Freidinger RM, Jurkiewicz NK, Lynch JJ, Salata JJ, Wang J, Dieckhaus CM, Slaughter DE, Vyas K (2003) Bioorg Med Chem Lett 13 (6):1165–1168
5. Haynes SW, Gao X, Tang Y, Walsh CT (2012) J Am Chem Soc 134(42):17444–17447
6. Breslin HJ, Kukla MJ, Kromis T, Cullis H, De Knaep F, Pauwels R, Andries K, De Clercq E, Janssen MAC, Janssen PAJ (1999) Bioorg Med Chem 7(11):2427–2436
7. Anzini M, Canullo L, Braile C, Cappelli A, Gallelli A, Vomero S, Menziani MC, De Benedetti PG, Rizzo M, Collina S, Azzolina O, Sbacchi M, Ghelardini C, Galeotti N (2003) J Med Chem 46(18):3853–3864

8. Chen CY, Lee PH, Lin YY, Yu WT, Hu WP, Hsu CC, Lin YT, Chang LS, Hsiao CT, Wang JJ, Chung MI (2013) Bioorg Med Chem Lett 23(24):6854–6859
9. Sandra CM, Cortes Eduardo C, Simon HO, Apan Teresa R, Camacho Antonio N, Lijanova V, Marcos MG (2012) Anti-Cancer Agents Med Chem 12(6):611–618
10. Sternbach LH (1979) J Med Chem 22(1):1–7
11. Merrifield RB (1963) J Am Chem Soc 85(14):2149–2154
12. Kamal A, Reddy KL, Devaiah V, Shankaraiah N, Reddy DR (2006) Mini Rev Med Chem 6 (1):53–69
13. Kaur N (2013) ChemInform 44(20)
14. Verma A, Ram Yadav M, Giridhar R, Prajapati N, Tripathi C, Saraf K (2013) Comb Chem High Throughput Screen 16(5):345–393
15. Bevacqua F, Basso A, Gitto R, Bradley M, Chimirri A (2001) Tetr Lett 42(43):7683–7685
16. Zhu Z, Sun ZY, Ye Y, McKittrick B, Greenlee W, Czarniecki M, Fawzi A, Zhang H, Lachowicz JE (2009) Bioorg Med Chem Lett 19(17):5218–5221
17. Combs AP, Saubern S, Rafalski M, Lam PY (1999) Tetr Lett 40(9):1623–1626
18. Ou L, Shao Z, Chen W, Giulianotti MA, Houghten RA, Yu Y (2011) Synlett 2011 (15):2259–2261
19. Pathak R, Roy AK, Kanojiya S, Batra S (2005) Tetr Lett 46(32):5289–5292
20. Bunin BA, Ellman JA (1992) J Am Chem Soc 114(27):10997–10998
21. Bunin BA, Plunkett MJ, Ellman JA (1994) Proc Natl Acad Sci U S A 91(11):4708–4712
22. Plunkett MJ, Ellman JA (1995) J Am Chem Soc 117(11):3306–3307
23. Plunkett MJ, Ellman JA (1995) J Org Chem 60(19):6006–6007
24. Plunkett MJ, Ellman JA (1997) J Org Chem 62(9):2885–2893
25. DeWitt SH, Kiely JS, Stankovic CJ, Schroeder MC, Cody DM, Pavia MR (1993) Proc Natl Acad Sci U S A 90(15):6909–6913
26. Dewitt SH, Schroeder MC, Stankovic CJ, Strode JE, Czarnik AW (1994) Drug Develop Res 33(2):116–124
27. Lattmann E, Billington DC, Arayarat P, Singh H, Offel M (1999) Sci Asia 25(2):107–112
28. Laustsen LS, Sams CK (2007) J Comb Chem 9(6):1094–1103
29. Lattmann E, Billington DC, Poyner DR, Arayarat P, Howitt SB, Lawrence S, Offel M (2002) Drug Des Discov 18(1):9–21
30. Matsushita K, Okamoto C, Yoshimoto M, Harada K, Kubo M, Fukuyama Y, Hioki H (2010) J Comb Chem 12(3):311–314
31. Lee K, Mun SJ, Ha DC, Lee KI, Gong YD, Lee JC (2006) Bull Korean Chem Soc 27 (7):974–975
32. Boutard N, Dufour-Gallant J, Deaudelin P, Lubell WD (2011) J Org Chem 76(11):4533–4545
33. Dufour-Gallant J, Chatenet D, Lubell WD (2015) J Med Chem 58(11):4624–4637
34. Bhalay G, Blaney P, Palmer VH, Baxter AD (1997) Tetr Lett 38(48):8375–8378
35. Zhang J, Goodloe WP, Lou B, Saneii H (2000) Molec Divers 5(3):127–130
36. Wu Z, Ercole F, FitzGerald M, Perera S, Riley P, Campbell R, Pham Y, Rea P, Sandanayake S, Mathieu MN, Bray AM, Ede NJ (2003) J Comb Chem 5(2):166–171
37. Hone ND, Wilson W, Reader JC (2003) Tetr Lett 44(46):8493–8495
38. Im I, Webb TR, Gong YD, Kim JI, Kim YC (2004) J Comb Chem 6(2):207–213
39. Lee JY, Im I, Webb TR, McGrath D, Song MR, Kim YC (2009) Bioorg Chem 37(3):90–95
40. Malik L, Kelly NM, Ma JN, Currier EA, Burstein ES, Olsson R (2009) Bioorg Med Chem Lett 19(6):1729–1732
41. Ali FE, Yuan CC, Ross ST, Hall LB (2000) Molec Divers 5(1):1–5
42. Thurston DE (1993) In: Molecular aspects of anticancer drug-DNA interactions. Springer, pp 54–88
43. Berry JM, Philip W, Thurston DE (2000) Tetr Lett 41(32):6171–6174
44. Kamal A, Reddy GSK, Raghavan S (2001) Bioorg Med Chem Lett 11(3):387–389
45. Kamal A, Reddy GSK, Reddy KL (2001) Tetr Lett 42(39):6969–6971
46. Kaplánek R, Krchňák V (2013) Tetr Lett 54(21):2600–2603

47. Shankaraiah N, Markandeya N, Espinoza-Moraga M, Arancibia C, Kamal A, Santos LS (2009) Synthesis 2009(13):2163–2170
48. Kamal A, Reddy GSK, Reddy KL, Raghavan S (2002) Tetr Lett 43(11):2103–2106
49. Kamal A, Reddy KL, Devaiah V, Reddy GSK (2003) Tetr Lett 44(25):4741–4745
50. Kamal A, Shankaraiah N, Reddy KL, Devaiah V (2006) Tetr Lett 47(25):4253–4257
51. Kamal A, Prabhakar S, Shankaraiah N, Markandeya N, Venkat Reddy P, Srinivasulu V, Sathish M (2013) Tetr Lett 54(33):4435–4441
52. Kamal A, Reddy KL, Devaiah V, Shankaraiah N, Reddy YN (2004) Tetr Lett 45 (41):7667–7669
53. Brouillette Y, Verdié P, Martinez J, Lisowski V (2008) Synlett 2008(15):2360–2364
54. Ludolph B, Waldmann H (2009) Tetr Lett 50(26):3148–3150
55. Kamal A, Reddy KL, Devaiah V, Shankaraiah N, Suresh Kumar Reddy G, Raghavan S (2007) J Comb Chem 9(1):29–42
56. Kamal A, Shankaraiah N, Devaiah V, Reddy KL (2006) Tetr Lett 47(37):6553–6556
57. Kamal A, Shankaraiah N, Prabhakar S, Reddy C, Markandeya N, Reddy KL, Devaiah V (2008) Bioorg Med Chem Lett 18(7):2434–2439
58. Lee SC, Park SB (2007) Chem Commun 36:3714–3716
59. Lemrová B, Soural M (2012) ACS Comb Sci 14(12):645–650
60. Fülöpová V, Gucký T, Grepl M, Soural M (2012) ACS Comb Sci 14(12):651–656
61. Fülöpová V, Krchňák V (2014) ACS Comb Sci 16(8):412–420
62. Trapani P, Volná T, Soural M (2016) ACS Comb Sci 18(6):349–354
63. Fülöpová V, Funk P, Popa I, McMaster C, Soural M (2015) Eur J Org Chem 2015 (16):3551–3557
64. Chatterjee N, Sarkar S, Pal R, Sen AK (2014) Tetr Lett 55(14):2261–2265
65. Nefzi A, Ong NA, Houghten RA (2001) Tetr Lett 42(31):5141–5143
66. Dadiboyena S, Arfaoui A, Amri H, Piedrafita FJ, Nefzi A (2015) Bioorg Med Chem Lett 25 (3):685–689
67. Schwarz MK, Tumelty D, Gallop MA (1998) Tetr Lett 39(46):8397–8400
68. Lee J, Gauthier D, Rivero RA (1999) J Org Chem 64(9):3060–3065
69. Wu CY, Sun CM (2002) Bioorg Med Chem Lett 12(6):959–962
70. Herpin TF, Van Kirk KG, Salvino JM, Yu ST, Labaudiniére RF (2000) J Comb Chem 2 (5):513–521
71. Kidwai M, Venkataramanan R (2004) Chem Heterocycl Compd 40(5):631–634
72. Kong KH, Chen Y, Ma X, Chui WK, Lam Y (2004) J Comb Chem 6(6):928–933
73. Sharma N, Chundawat TS, Mohapatra SC, Bhagat S (2013) RSC Adv 3(37):16336–16339
74. Goff DA, Zuckermann RN (1995) J Org Chem 60(18):5744–5745
75. Verdié P, Subra G, Averland-Petit MC, Amblard M, Martinez J (2008) J Comb Chem 10 (6):869–874
76. Verdié P, Subra G, Feliu L, Sanchez P, Bergé G, Garcin G, Martinez J (2007) J Comb Chem 9 (2):254–262
77. Mayer JP, Zhang J, Bjergarde K, Lenz DM, Gaudino JJ (1996) Tetr Lett 37(45):8081–8084
78. Gerdes JM, Waldmann H (2003) J Comb Chem 5(6):814–820
79. Migihashi C, Sato F (2003) J Heterocycl Chem 40(1):143–147
80. Hulme C, Ma L, Kumar NV, Krolikowski PH, Allen AC, Labaudiniere R (2000) Tetr Lett 41 (10):1509–1514
81. Smith RA, Bobko MA, Lee W (1998) Bioorg Med Chem Lett 8(17):2369–2374
82. Hulme C, Peng J, Morton G, Salvino JM, Herpin T, Labaudiniere R (1998) Tetr Lett 39 (40):7227–7230
83. Chen JJ, Golebiowski A, Klopfenstein SR, West L (2002) Tetr Lett 43(22):4083–4085
84. Kennedy AL, Fryer AM, Josey JA (2002) Org Lett 4(7):1167–1170
85. Jeon MK, Kwon JJ, Kim MS, Gong YD (2008) Synlett 2008(11):1651–1656
86. Boojamra CG, Burow KM, Ellman JA (1995) J Org Chem 60(18):5742–5743
87. Ettmayer P, Chloupek S, Weigand K (2003) J Comb Chem 5(3):253–259

88. Joseph CG, Wilson KR, Wood MS, Sorenson NB, Phan DV, Xiang Z, Witek RM, Haskell-Luevano C (2008) J Med Chem 51(5):1423–1431
89. Rivero IA, Peralta M, Heredia S, Madrigal D, Pina-Luis G, Chávez D (2003) Arkivoc 2003 (11):27–36
90. Nefzi A, Ostresh JM, Houghten RA (1997) Tetr Lett 38(28):4943–4946
91. Nefzi A, Dooley C, Ostresh JM, Houghten RA (1998) Bioorg Med Chem Lett 8 (17):2273–2278
92. Moure A, Sanclimens G, Bujons J, Masip I, Alvarez-Larena A, Pérez-Payá E, Alfonso I, Messeguer A (2011) Chem Eur J 17(28):7927–7939
93. Krchňák V, Weichsel AS (1997) Tetr Lett 38(42):7299–7302
94. Giovannoni J, Subra G, Amblard M, Martinez J (2001) Tetr Lett 42(32):5389–5392
95. Caroen J, Clemmen A, Kámán J, Backaert F, Goeman JL, Fülöp F, Van der Eycken J (2016) Tetrahedron 72(1):148–160
96. Sülli-Vargha H, Schlosser G, Ilaš J (2007) J Peptide Sci 13(11):742–748
97. Lampariello LR, Piras D, Rodriquez M, Taddei M (2003) J Org Chem 68(20):7893–7895
98. Xu JF, Huang X (2009) J Comb Chem 11(5):938–942
99. Cao J, Huang X (2010) J Comb Chem 12(1):1–4
100. Huang X, Cao J, Huang J (2009) J Comb Chem 11(4):515–518
101. Olsen CA, Christensen C, Nielsen B, Farah MM, Witt M, Clausen RP, Kristensen JL, Franzyk H, Jaroszewski JW (2006) Org Lett 8(15):3371–3374
102. Ottesen LK, Olsen CA, Witt M, Jaroszewski JW, Franzyk H (2009) Chem Eur J 15 (12):2966–2978
103. Li D, Hall DG (2005) Tetrahedron: Asymmetry 16(10):1733–1736
104. Eans SO, Ganno ML, Mizrachi E, Houghten RA, Dooley CT, McLaughlin JP, Nefzi A (2015) J Med Chem 58(12):4905–4917
105. Vaňková B, Brulíková L, Wu B, Krchňák V (2012) Eur J Org Chem 2012(26):5075–5084
106. Ventosa-Andrés P, Barea Ripoll CA, La-Venia A, Krchňák V (2015) Tetr Lett 56 (40):5424–5428

Top Heterocycl Chem (2017) 52: 65–94
DOI: 10.1007/7081_2016_1

Published online: 23 December 2016

Intramolecular N-Acyliminium Cascade (INAIC) Reactions in Cyclization of Peptide-Like Molecules

Frederik Diness, Yuanyuan Wang, and Morten Meldal

Abstract The present review describes the implementation of *N*-acyliminium ion chemistry of peptide aldehydes in the context of solid phase peptide synthesis. The INAIC reaction is a cascade reaction in which an aldehyde, located within a peptide, initially reacts with proximate and weakly nucleophilic amide nitrogen to form a hydroxylactam. Under acidic conditions the hydroxylactam eliminates water and produces highly reactive *N*-acyliminium ions. These in turn reacts with high stereospecificity with any nucleophile nearby in the peptide, including C-nucleophiles or even a second amide nitrogen to form two new heterocyclic rings which can both be 5 membered or they can be 5,6; 6,5; 6,6; or 5,7 membered. In this manner simple peptides can be transformed into interesting heterocycles with structures that may interact with 7TM receptors and are valuable in drug screening programs. In addition to amides, carbamoyl nitrogens can also act as the primary nucleophile in the analogous INCIC reaction, thus expanding the scope of these reactions significantly. Due to the high reactivity of the *N*-acyl-iminium ions deactivated C-nucleophiles such as dichlorobenzene rings may be employed in addition to the reactive C-nucleophiles, e.g., indoles commonly used in the related Pictet-Spengler reaction. In this manner complex annulated tetrahydro β-carbolines and tetrahydroisoquinoline can be synthesized within the peptide framework. The scope of the INAIC and INCIC reactions is significant with more than 40 different heterocyclic scaffolds currently synthesized and even more new scaffolds possible.

Keywords Cascade reaction • Complex heterocycles • Intramolecular cyclization • *N*-acyliminium ion chemistry • Peptide aldehyde • Peptide alkaloids • Peptide cyclization • Solid phase INAIC reaction

---

F. Diness, Y. Wang, and M. Meldal (✉)
Department of Chemistry, CECB, Universitetsparken 5, 2100 Copenhagen, Denmark
e-mail: meldal@nano.ku.dk

Contents

1 Introduction

Ever since the introduction of the CuAAC click chemistry [1–3] there has been an ongoing quest to find other reactions that provide the properties of reactivity and more importantly selectivity required for a click reaction [4, 5]. Selectivity may be acquired through specific interaction and activation by catalysis as in CuAAC or it may simply be achieved through proximity driven reactivity enhancement. This is the principle behind the intramolecular *N*-acyliminium cascade reactions (INAIC), a generic set of reactions performing as indicated in Scheme 1. The INAIC reactions are initiated from **1** by formation of a strong electrophile **2,** e.g. an aldehyde in the vicinity of a weak nucleophile that upon intramolecular nucleophilic attack on the electrophile provides a second electrophile, **3**. This can then react with a second weak nucleophile and cyclize into complex heterocycles **4**. Since the reaction is intramolecular, it is ideally suited for solid phase chemistry, which is the focus of the present review.

The solid phase INAIC reaction is one out of a range of reactions taking advantage of the reactivity of imines and iminium ions towards nucleophiles, including the Mannich reaction with the α-carbon nucleophile [6], the Petasis reaction with aryl/vinylboronic acids [7] and the Pictet–Spengler reaction with nucleophilic aromatic compounds [8]. Although all these reactions follow a pattern similar to that outlined in Scheme 1 of a nucleophilic attack on a preformed imine or iminium ion, they are considered distinct reaction types. The INAIC reaction is different from the Pictet–Spengler reaction (Scheme 2), which employs amine nucleophile **5** as Nu_1 (Scheme 1), in that the first nucleophile Nu_1 of INAIC reactions is a very weak amide [9] or carbamoyl nitrogen [10] (INCIC-reaction) (Scheme 3). This provides many opportunities for use of the amide rich peptide framework in INAIC and INCIC reactions.

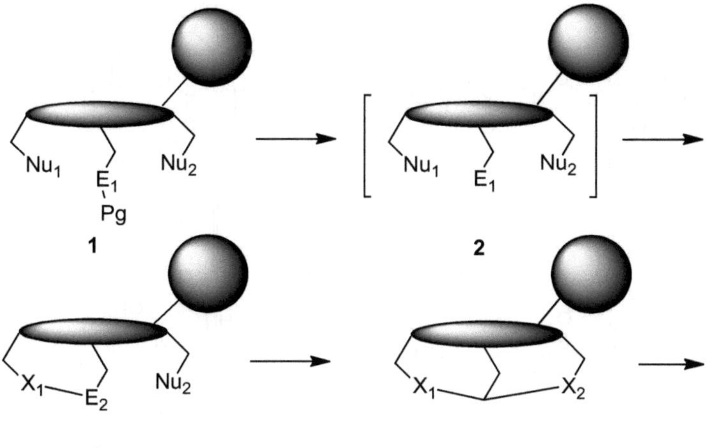

3 4

Scheme 1 The solid phase *INAIC* reaction provides complex heterocycles from easily available peptide-like structures through two successive intramolecular reactions of a reactive electrophile, mostly an aldehyde or ketone, with two proximate nucleophiles where Nu_1 is an amide bond and Nu_2 can be a C, N, O or S nucleophile including a second amide bond

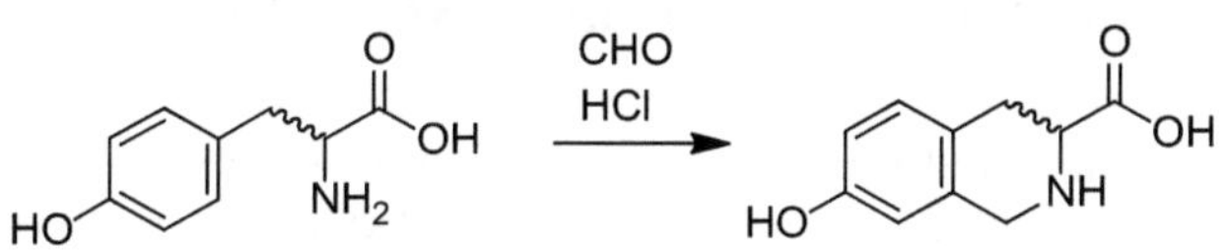

5 6

Scheme 2 In the original Pictet–Spengler reaction of tyrosine with formaldehyde under acidic conditions in 1911 [8], the aldehyde forms an imine, which under the acidic conditions is subject to nucleophilic attack by the adjacent aromatic ring

Intramolecular *solution* phase *N*-acyliminium cyclizations were described in an excellent review by Maryanoff et al. in 2004 [12] and the Pictet–Spengler reaction was reviewed more generally by Cox and Cook in 1995 [13]. Solid phase INAIC reactions were also previously described in brief reviews [14–17].

The present review concerns the performance of *N*-acyliminium cyclization reactions on solid support and we will compare these strategically to similar solution phase chemistry and selected Pictet–Spengler reactions. Our contribution to the *N*-acyliminium chemistry on solid phase was initiated by the observation that N-terminal aldehydes **8**, obtained by periodate oxidation of N-terminal serine, e.g. **7**, in resin bound peptides in addition to expected formation of their hydrates **9** also formed internal adduct **10** with backbone amides in the preceding peptide chain (Scheme 3) [11]. Under hydrolytic conditions this even leads to peptide cleavage [18].

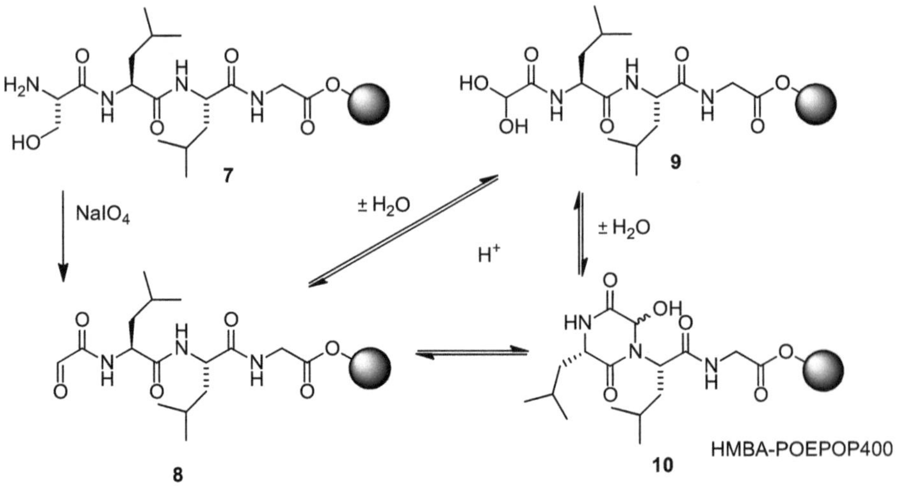

Scheme 3 MAS-NMR of a peptide aldehyde obtained from oxidation of an N-terminal serine of Ser-Leu-Leu-Gly-HMBA-POEPOP$_{400}$ provided evidence that the hydroxylactam **10** forms in equilibrium with the hydrate **9** even under neutral and basic conditions on solid support [11]

2 Mechanistic Aspects INAIC and INCIC Reactions

The entropically favoured formation of a cyclic transition state of these intramolecular tandem cyclization reactions (INAIC and INCIC) in combination with the low reactivity of the amide nitrogen as a nucleophile ensures the high selectivity for a five- or six-membered iminium ion ring. When the electrophile is an aldehyde or a ketone the first reaction obviously proceeds through formation of a hydroxylactam, e.g. **12**, which depending on the conditions in a second step loose water to provide the highly reactive *N*-acyliminium ion **13** that reacts with high stereo-specificity to give **14** exclusively. The second nucleophile (thiophene in Scheme 4) can be any reactive electrons in the vicinity of the formed iminium ion, including those of amines, alcohols, thiols, amides, acetylides and π-electrons of a large variety of aromatic residues [12].

The initial attack of the amide bond on the carbonyl electrophile and thereby the ability to provide a hydroxylactam is crucial for the success of the INAIC-reaction. This is highly favoured by the ability to form a 5- or 6-membered transition state towards **12**. Once the stereoisomeric mixture of hydroxylactams has been formed these may actually be isolated and constitute relatively stable intermediates of the reaction [19]. The clean formation of hydroxylactams is favoured by intermediate acidity, e.g. 10% of TFA in water. However, if the second nucleophile is very reactive, even minute conversion to iminium ion **13** may lead to formation of the final product. The transformation of the two stereoisomeric hydroxylactams to the *N*-acyliminium ion is better catalysed by stronger non-aqueous acids [19]. According to *ab initio* calculations the charge distribution of the iminium

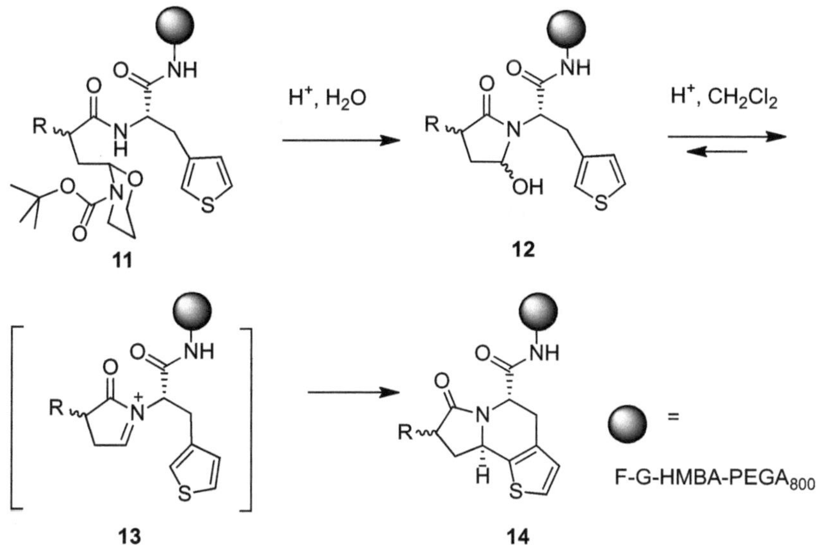

Scheme 4 Example of a typical INAIC reaction. The aldehyde in the form of a protected building block attached to the N-terminal of a peptide (**11**) containing a penultimate 3-thienylalanine is rapidly released with acid to form the hydroxylactam **12**. The subsequently elimination of water under anhydrous conditions provides the N-acyliminium ion **13**, which reacts with the thiophene acting as a C-nucleophile to provide **14**

ion (Fig. 1) is mainly located on the nitrogen with significant charge distributed to the iminium carbon. The attack of the second aryl nucleophile on the *N*-acyliminium ion can proceed either through initial formation of a five- or a six-membered cyclic intermediate involving either one of two aryl carbons as dictated by the steric requirements, the reactivity of the iminium electrophile, electron density on the two competing carbon atoms and other stereo-electronic factors [13]. If formed, the five-membered spiroindoline intermediate rapidly rearranges to the more stable six-membered ring of the reaction product [20]. The more reactive the electrophile, the more the reaction occur by direct attack to give the 6-membered ring [20]. In the special case of indole of tryptophan the charged five-membered intermediate is stabilized. It may form kinetically and subsequently immediately rearrange to the more stable six-membered intermediate.

Generally speaking, during formation of the transition state for the attack of Nu_2, the π-electrons of the iminium ion locate on the nitrogen and the positive charge resides mainly on the attacking aryl carbon. The carbonium ion intermediate then loses an aryl proton and rearranges to the neutral product. Overall, this would then be equivalent to a Friedel-Crafts type of reaction.

The INAIC and INCIC reactions with less reactive C-nucleophiles only perform in completely anhydrous acids, e.g. TFA/DCM or formic acid and it may be suggested that the actual formation of the iminium ions from the hydroxylactams is facilitated through acylation by small amounts of the acid anhydride present in the anhydrous

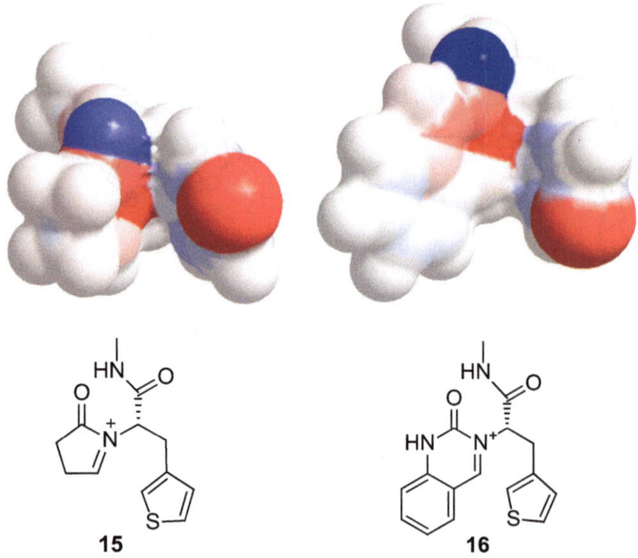

Fig. 1 The electron densities were computed for compounds **15** and **16** by a Hartree–Fock calculation using the Pople 3-21G basis set and performed in the Gamess interface (Charge, *blue:* negative, *red:* positive). The positive charge on the iminium carbon of *N*-acyliminium ion is larger than that of the *N*-carbamoyliminium ion. The electron density is high at the reacting 2-position of the thiophene ring creating a perfect scenario for bond formation

acid under such conditions. In fact the reaction performs particularly well under aprotic conditions of TFA anhydride, BF_3 and pentamethyl piperidine in DCM (Unpublished data). A similar dehydration effect may be the actual cause of activity in the recently described cyanuric chloride catalysis of Pictet–Spengler reactions [21]. When Nu_2 is a very reactive C-nucleophile, ring closure is also observed under neutral Dean-Stark conditions of forced formation of an imine in refluxing toluene [12], indicating that with protected aldehyde building blocks the strong acid is mainly required for conversion of *N,O*-acetal or *O,O*-acetal to imine [22].

For *N*-acyliminium ions the stereochemistry of the product is entirely *trans* over the C-N-C bond of the second ring formed according to careful analysis by LC-MS and 2D-NOESY ^{1}H-NMR spectroscopy. The stereochemistry is therefore fully controlled by the chirality of the amino acid carrying the nitrogen of the *N*-acyliminium ion.

The nucleophilic character and proximity of other functional groups that can act as alternative nucleophiles is quite important for the outcome of the reaction. For example, when the reaction was performed in solution with the free carboxylate of the amino acid carrying the iminium nitrogen, this acted as a competitive nucleophile providing **18** instead of **17**. When the intended nucleophile Nu_2 is very unreactive the subsequent amide bond could also react to give **19** (Fig. 2) [23].

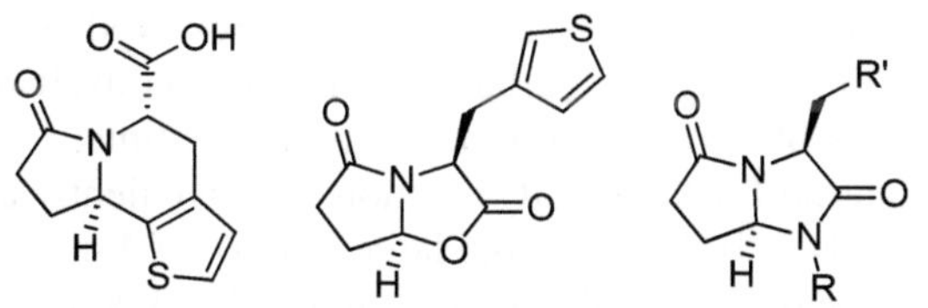

17 18 19

Fig. 2 The reactivity of the *N*-acyliminium ion is high and once formed, in absence of a reactive C-nucleophile, it will eventually find another appropriate nucleophile in the surrounding molecular framework, be it an adventurous carboxylate or even another backbone amide

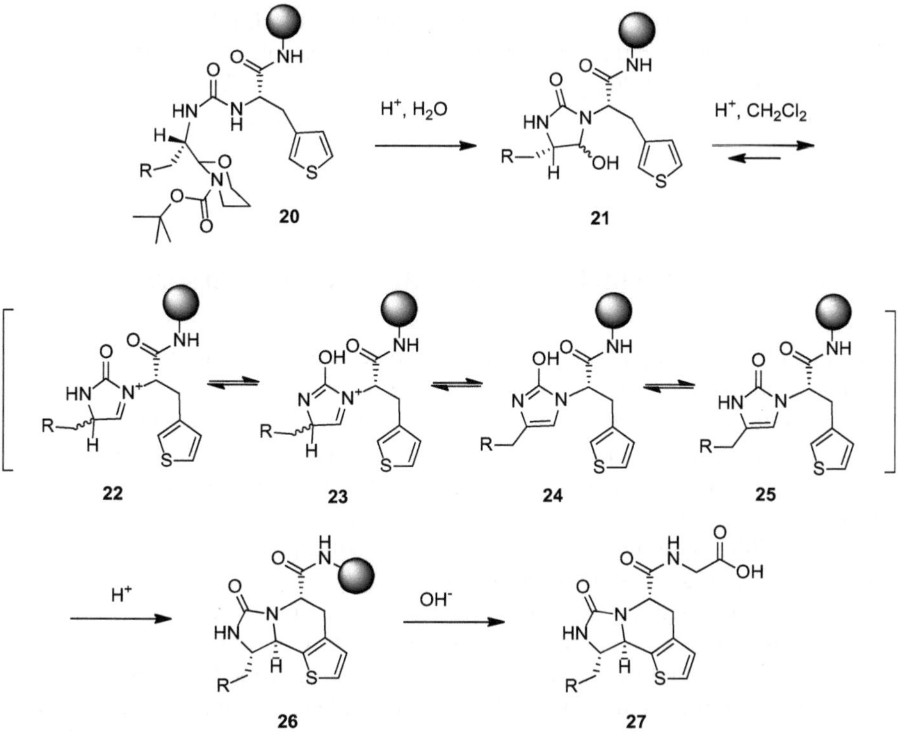

Scheme 5 The INCIC reaction employs the reactive carbamoyliminium ions for reactions with nucleophiles. The building blocks for this reaction are readily available from amino acids. The chirality may be partially lost due to the equilibrium involving loss of an α-proton

The solid phase INCIC reactions [10] are the equivalent of the INAIC reaction in which a carbamoyliminium ion intermediate is involved (e.g. **22** in Scheme 5). Generally, the electrophile of the *N*-acyliminium ions is thought to be less reactive than the carbamoyliminium ions due to the electron withdrawing effect of the additional heteroatom [12] This is in contrast to observed reactivity of amides vs. urethanes [24] Frequently, the carbamoyliminium ion itself is very short lived and it is rarely accessible over any significant period of time.

The starting material for the INCIC reaction is conveniently all the natural and unnatural amino acids [25]. Notably, the large variety of *N*-protected amino acids available today can be transformed into protected amino aldehydes via the Weinreb amide reduction and coupled to the N-terminal of a peptide by a urea bond formation, e.g. **20** which upon treatment with acid gives **21**. The caveat of the INCIC reaction is that five-membered carbamoyliminium ions **22** effectively are equivalents to imidazolones **23** and **24** and therefore readily isomerize to the enamides when a proton is present at the α-carbon neighbouring the iminium ion. The equilibrium is in favour of the enamide **24** and the presence of the enamide provides a much less reactive electrophile with loss of chirality at the α-carbon. However, depending on the structure it may still be possible to convert the enamide to the cyclized product **26** under strongly acidic conditions [26].

In six-membered *N*-carbamoyliminium ions α-protons are less acidic and enamides do not form.

3 Formation of Aldehyde from Aldehyde Precursors

In contrast to solution phase Pictet–Spengler reactions where the free aldehyde can mostly be added, the INAIC reactions require protected aldehyde building blocks during assembly of the precursor for the cyclization cascade reaction. The source for in situ formation of the carbonyl electrophile can be quite important for the outcome of the reaction and in particular the compatibility with other chemistry in the target compound, often an issue that is crucial for successful solid phase chemistry. Other requirements for a successful solid phase reaction are that it is clean, fast, quantitative and homogeneous.

A number of reactions are available for converting other functional groups into aldehyde or ketone on solid support, see Scheme 6. C-terminal aldehydes can be obtained by controlled $LiAlH_4$ reduction of N-methoxy N-methyl carboxamides **28** [27, 28] by reduction or alkylation of succinimide **29** [12] or by controlled oxidation of alcohols, e.g. **42–44** with periodane to ketone [29] or aldehyde [30, 31]. Vicinal di-hydroxyl compounds from e.g. **31** and β-hydroxyl amines **32** may be cleaved oxidatively by IO_4^- to give aldehyde.

This reaction [32] originally introduced to oxidize peptides with an N-terminal serine [33] has been very popular [11, 34, 35]. Double bonds is a versatile source of reactive aldehydes and conversion from alkene to aldehyde may be effected either by ozonolysis [36] or more conveniently by optimized conditions of catalytic OsO_4 dihydroxylation and instantaneous oxidative cleavage with IO_4^- in presence of DABCO [30, 37].

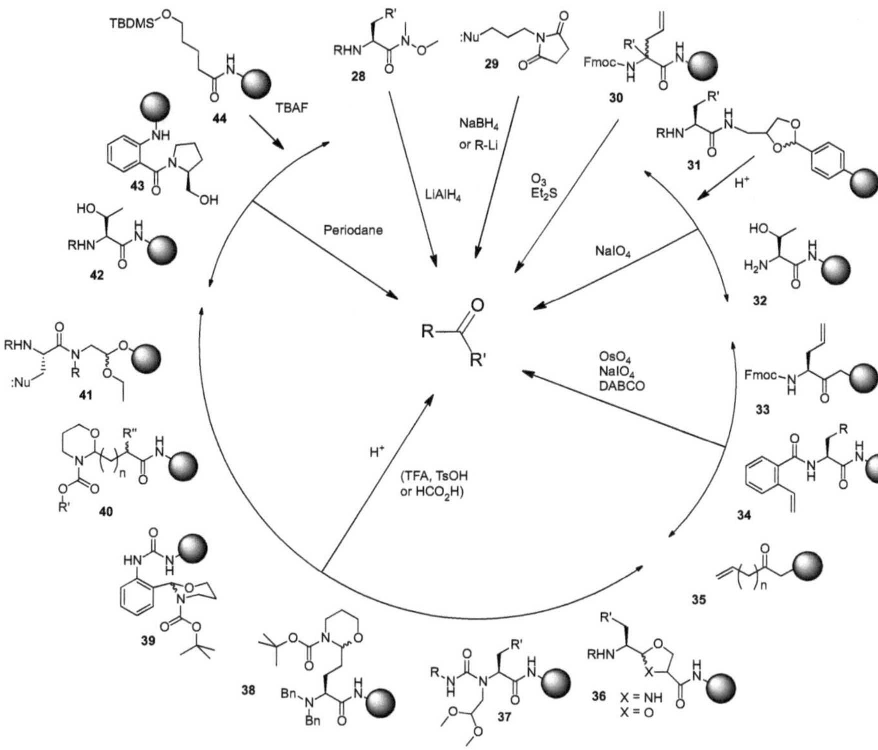

Scheme 6 The method used of aldehyde generation is crucial for application perspective and reaction outcome in INAIC and INCIC reactions. Methods include reduction, oxidation, oxidative cleavage, rearrangements and acetal cleavage, dihydroxylation/oxidation and more

Alternatively, an approach may be employed in which a building block containing an appropriately protected aldehyde for easy and orthogonal release of the reactive aldehyde on support is first synthesized and then this is incorporated during assembly of peptides containing this aldehyde precursor. In this regard, the Boc-protected *N,O*-acetals of the aldehyde (Box-group) [25, 38] have proven particularly useful and allowed manipulation with functional groups to facilitate clean solid phase synthesis with aldehydes [9, 10, 19, 22, 23, 30]. Upon rapid release of the aldehyde at the right place and time, almost quantitative conversions to single products may be achieved in this manner.

There are in principle other methods that can generate iminium ions or will lead to formation of the same type of products. The CAAPS: Coupling-Amination-aza-Annulation-Pictet-Spengler reaction sequence [39] Scheme 7a and Ru catalysed isomerization of allylic amides, **46** [40] Scheme 7b gives rise to formation of reactive *N*-acyliminium ions **45** and **47** equivalent to those generated from aldehyde. The combination of Ru catalysed metathesis with Ru catalysed rearrangement of allylic amide to *N*-acyliminium ion **50**, Scheme 8, as precursor

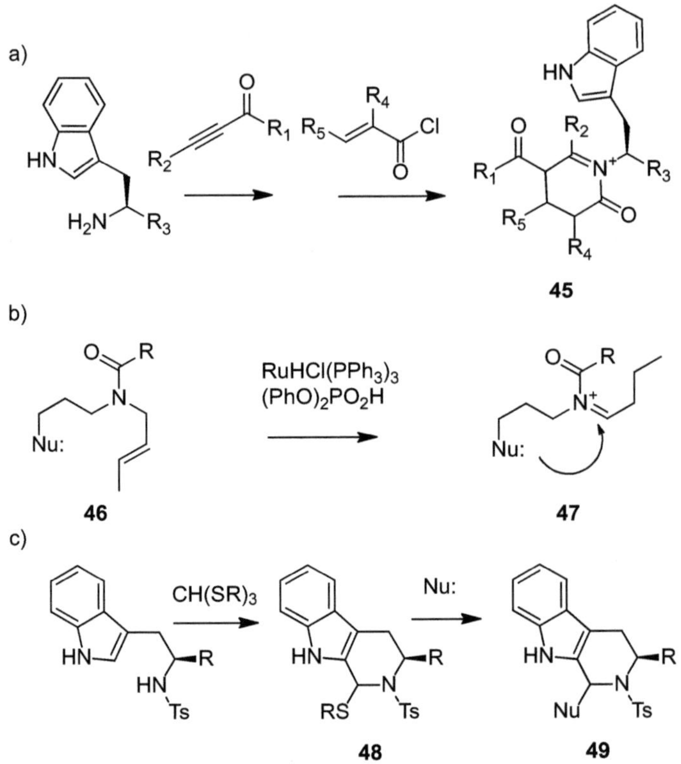

Scheme 7 Alternative routes to *N*-acyliminium ions and β−carbolines. a) A multicomponent cyclization with concomitant annulation end rearrangement of an enamide to an *N*-acyliminium ion **45**. b) Ru catalysed rearrangement of allylic amides **46** to *N*-acyliminium ion **47**. c) A method involving three successive nucleophilic substitutions on thio-orthoformate and using a variety of nucleophiles in the last step generating tetrahydro β-carbolines, **49**

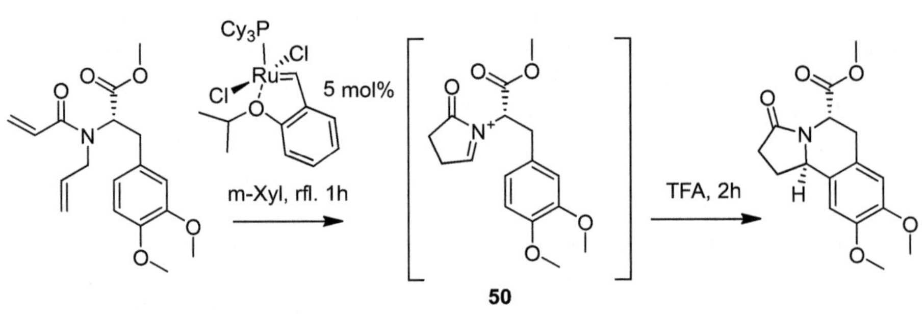

Scheme 8 The combination of Ru catalysed metathesis and Ru catalysed isomerization of the allylic amide formed to *N*-acyliminium ion is a powerful method for the generation of diversity in the INAIC reactions, which is independent of the ring closure of hydroxylactam formation

to INIAC reaction, is particularly elegant [41]. Enamides, the initial rearrangement product in this process are also frequently used acyliminium precursors in solution INAIC reactions [12]. The third method is different and involve double nucleophilic attack on thio-orthoformate with e.g. *N*-tosyl-tryptamine to install an *S,N*-acetal with a thiolate leaving group in **48** prone to attack with further nucleophiles giving **49** (Scheme 7c) [42].

Of these methods only the Ru catalysed rearrangement has been demonstrated on solid support. Two of the methods for generation of aldehyde have been particularly useful and will be described in more detail below.

4 Aldehyde from Protected *N,O*-acetals and Dihydroxylation/Oxidation of Alkenes

The protection of aldehyde building blocks with 3-aminopropanol followed by locking the formed cyclic *N,O*-acetal by installing a Boc-group provided a stable aldehyde protection, e.g. in **51**, that could rapidly release the aldehyde on solid support. The cleavage requires hydrolytic conditions and in 10% aqueous TFA the hydroxylactam forms immediately upon release of the aldehyde. In this regard the nitrogen in the *N,O*-acetal is important because protonation is immediate and quantitative upon cleavage of Boc with the acid. With proximate and reactive C-nucleophiles such as the indole, **51**, this proceeds almost instantaneously to the INAIC product **52** with a large variety of substitutions on the indole ring, Scheme 9 [9].

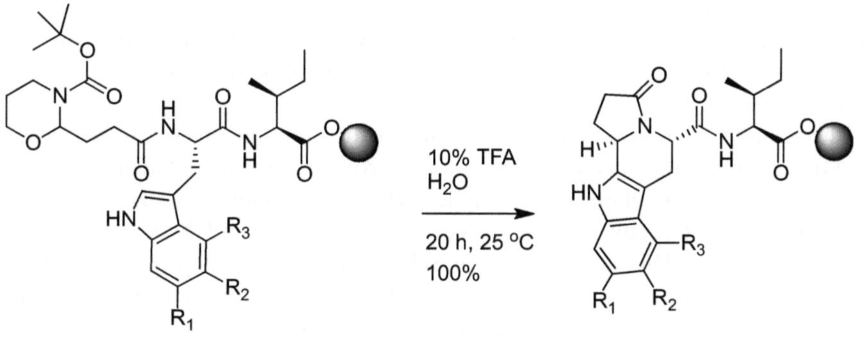

51 **52**

Scheme 9 Solid phase INAIC chemistry was presented by synthesis of a large variety of substituted tetrahydro β-carbolines

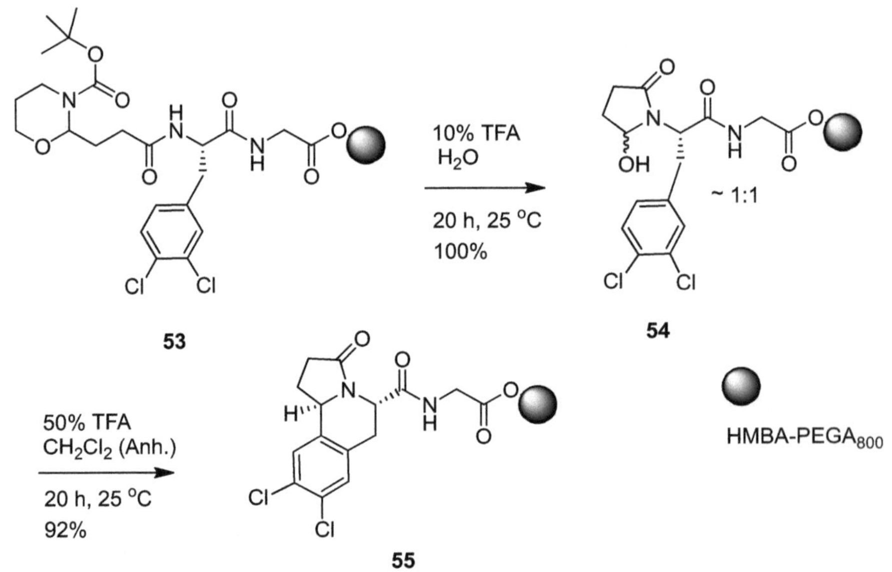

Scheme 10 Even highly deactivated C-aryl nucleophiles can be brought to react quantitatively in an INAIC reaction. This requires a two-step procedure where the hydroxylactam is first formed under hydrolytic conditions followed by more forcing conditions of anhydrous TFA to form significant amounts of the N-acyliminium ion

However, less reactive nucleophiles, as in **53**, require anhydrous acids as mentioned above and here the solid support allows rapid and convenient exchange of conditions. This two-step procedure, via the hydroxylactam **54** as presented in Scheme 10, provides a quantitative conversion to the INAIC product, e.g. **55** in a completely stereo-selective manner and without formation of by-products. The direct conversion with strong acid is less successful. Recently the analogous Fmox-protection was described as a base labile alternative to the Box-protection of aldehydes [43].

The catalytic dihydroxylation of alkenes, with simultaneous oxidative cleavage and regeneration of catalyst using periodate, provides a clean and quantitative solid phase conversion of alkenes to aldehydes. The diverse source of alkene containing building blocks thereby becomes a very rich source of aldehyde precursors for INAIC reactions. This is illustrated in Scheme 11 where a peptide **56** terminated with 2-vinyl benzoic amide is converted to the cyclization product **57** in two steps. It is important to note that the solid support in this case facilitates the rapid isolation of the intermediately formed hydroxylactam from the basic oxidation mixture prior to acid treatment. Once the hydroxylactam is formed, acid treatment gives almost quantitative product formation [30].

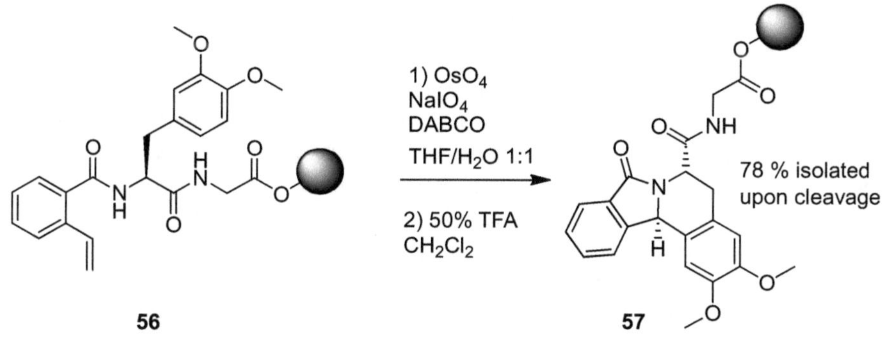

56 57

Scheme 11 Solid phase dihydroxylation in situ oxidation of vinyl groups with a catalytic amount of OsO_4 and $NaIO_4$ and DABCO gives an almost quantitative formation of the peptide hydroxylactam

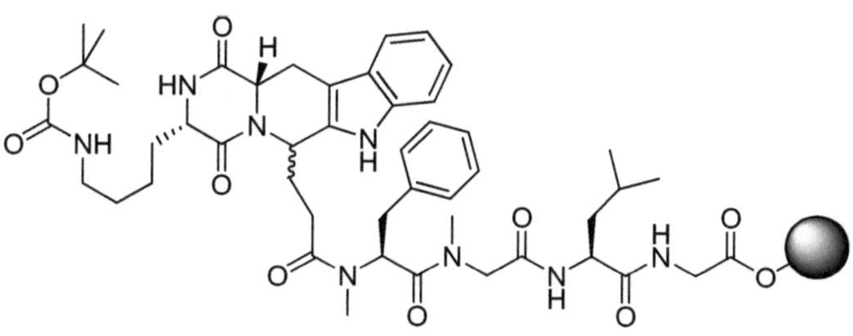

58

Fig. 3 The intermolecular reaction on N-terminal peptide aldehydes requires protection of the amine at the two penultimate amide bonds to prohibit formation of peptide N-acyliminium ions. Upon Pictet–Spengler reaction with tryptophan methyl ester and coupling of amino acid to the tetrahydro-β-carboline, a fused diketopiperazine, e.g. **58**, can be synthesized in quantitative yield on solid support

5 N-Terminally Located Aldehydes on Solid Phase

Since the reaction to form hydroxyl lactams with backbone amides is almost instantaneous the use of the peptide bound aldehydes for reaction with external nucleophiles requires backbone amide protection.

N-Methylation of one or better two of the penultimate amino acids was found to be a safe method to obtain N-terminal peptide aldehydes for a variety of external manipulations on solid support including Pictet–Spengler reactions to produce e.g. compound **58** upon subsequent diketopiperazine formation through ester aminolysis (Fig. 3) [18, 38]. The formation of tetrahydro-β-carboline on solid support also provided the opportunity to continue peptide synthesis from the tetrahydro-β-carboline nitrogen to incorporate this structure internally in the peptide chain [18].

6 Aldehydes in Peptide Sidechains

The dihydroxylation/oxidative cleavage reaction described above could also be used on protected peptides e.g. **59** on solid support as demonstrated during synthesis of the constrained enkephalin analogue **61** [30]. In this approach the INAIC reaction is performed with an aldehyde located in amino acid sidechain and generated from the readily available allylglycine building block as a source of side chain aldehydes by clean oxidative cleavage with OsO_4/IO_4^-/DABCO. The synthesis was performed with the same outcome by performing the INAIC reaction during or subsequent to assembly (Scheme 12) of the enkephalin peptide **60**.

In an analogues fashion a conventional Pictet–Spengler reaction was performed using an α-alkylallylglycine building block and O_3/Et_2S for the oxidative cleavage of the alkene. The quaternary α-allylphenylalanine was conveniently installed on solid support by allylation of aryliminophenylalanines. The Pictet–Spengler reaction of **62** with tryptamine or dopamine occurred to give e.g. **63** with subsequent cyclative cleavage from the solid support to tri- or tetra-cyclic spiro compounds **64**. The cyclative cleavage discriminated products from by-products provided very pure mixtures of the four diastereomeric products **64** formed in the reaction, Scheme 13, and a crystal structure was obtained from both reactions upon diastereoisomer separation [36].

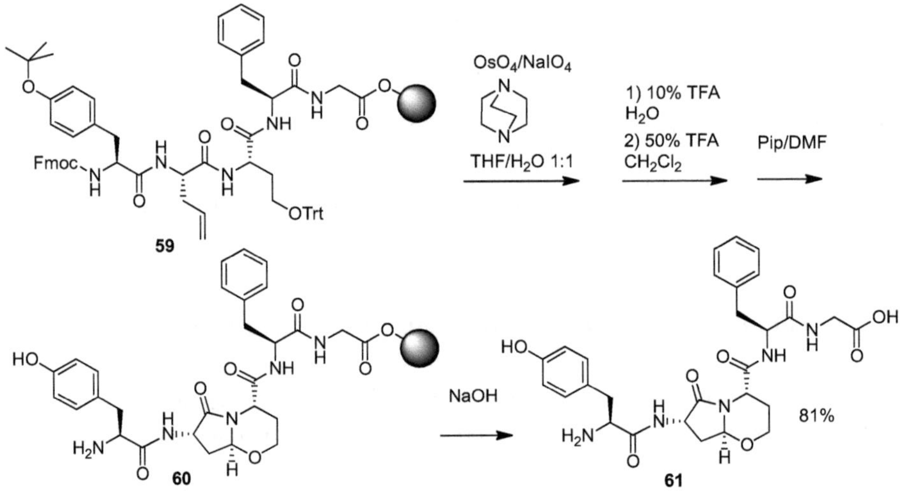

Scheme 12 Post assembly solid phase INAIC reactions (e.g. **59**) may serve to produce constrained analogues of bioactive peptides, e.g. enkephalin **60** in high yield and purity using oxidative cleavage of centrally located allyl glycine. Notably, the aldehyde only reacts to selectively form a five-membered ring with the preceding amino acid nitrogen

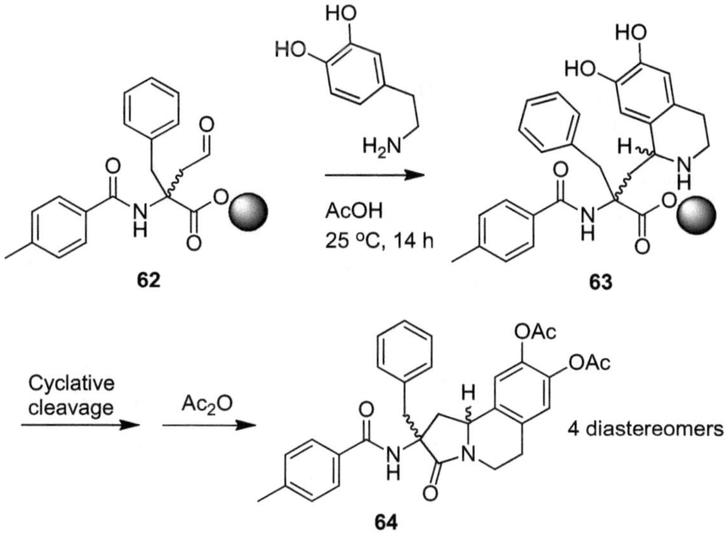

Scheme 13 Pictet–Spengler reaction of side chain aldehyde produced by α-allylation of amino acids and oxidative alkene cleavage with tryptamine and dopamine produced the diastereomeric mixture of products **64** upon cyclative cleavage form the resin

Allyl groups and acetals as aldehyde precursors can also be introduced by simple peptoid alkylations during assembly of a scaffold for INAIC reactions. This in turn gives access to a different class of molecular templates [44].

7 C-Terminal Peptide Aldehydes

An elegant approach of using C-terminal peptide aldehydes in conventional Pictet–Spengler reactions to ligate two larger peptide structures **65** and **66** via a tetrahydro-*β*-carboline **67** has been reported by the group of Tam. The method was compared with generating the tetrahydro-*β*-carboline with Fmoc-aminopropanal followed by stepwise peptide assembly. The direct ligation with peptide aldehydes compared well with the assembly of **68**, subsequent to Pictet–Spengler reaction, Scheme 14 [45]. The precursor peptide aldehydes were assembled as 2,3-dihydroxypropylamides linked to the support as a benzylacetal that upon TFA catalysed hydrolysis and periodate oxidation provided the C-terminal peptide aldehyde, e.g. **66** [34].

C-terminal aldehydes were also produced from resin bound acetal **69**. Upon release of the aldehyde this reacted with an amide or urethane bond on the subsequent amino acid to give a bridged bicyclic ring system **70** with the interesting 3-dimentional topology of e.g. **71** presented in Scheme 15 [24].

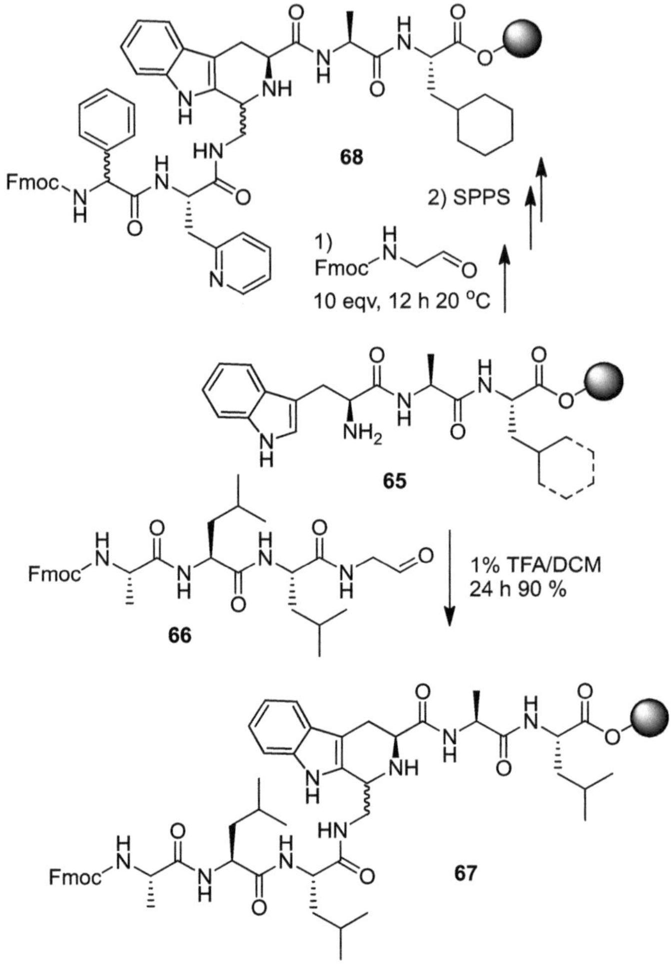

Scheme 14 Elegant peptide ligation via formation of tetrahydro-β-carboline between an N-terminal tryptophan and a C-terminal peptide aldehyde. Yields and purity compared well with stepwise assembly

8 Bridge Head Substitution

The use of ketones as the first electrophile significantly influences the reactivity in both the first and the second steps of the INAIC reaction and cyclization only occurs in particularly favourable configurations and conditions. While several reports have appeared on N-acyliminium ion reactions with ketone electrophiles in solution [12] this reaction on solid support has been significantly more difficult. Nielsen et al. scanned a number of catalyst systems for the successful cyclization and found that as observed in solution, anhydrous formic acid [12, 44] was particularly

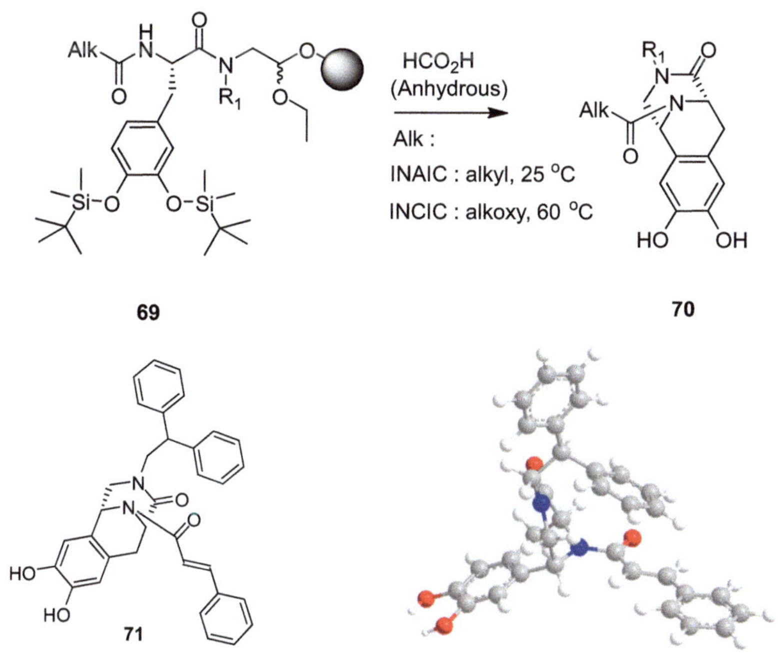

Scheme 15 Release of C-terminal aldehyde and subsequent INAIC cyclization in solution provided topologically interesting scaffolds

well suited for this solid phase reaction of ketones, e.g. as in the conversion of **72** via iminium ion **74** to bridgehead substituted **75**, Scheme 16 [46]. The stereoisomer **76** resulting from cis nucleophilic attack was not observed. The ease of initial formation of the hydroxylactam is the first essential step in the INAIC reactions and this seems to occur even with ketones in presence of most acids. However, the subsequent elimination of water and reaction with nucleophile may be hindered by the presence of a bridgehead substitution.

In another study threonines in **77** were quantitatively oxidized by periodane to provide ketones. This in turn formed cyclic Boc-imine by reaction with a protected α-amine in the subsequent amino acid. However, the six-membered iminium ion ring formed, rapidly rearranged to the stable pyrazinones **78** prior to any nucleophilic attack in agreement with the above observations of steric bulk in the transition state of the nucleophilic attack on ketone derived *N*-acyliminium ions (Scheme 17) [29].

While the *N*-acyliminium ion of the INAIC reaction is sterically hindered in reaction with ketones this seems to be less of a problem in the conventional Pictet–Spengler reaction (Scheme 18) where a ketone reacts with a primary amine, **79** and the protonated imine subsequently reacts with a second nucleophile giving, e.g. **80**. The reaction was performed on a PEG support as the ester for subsequent

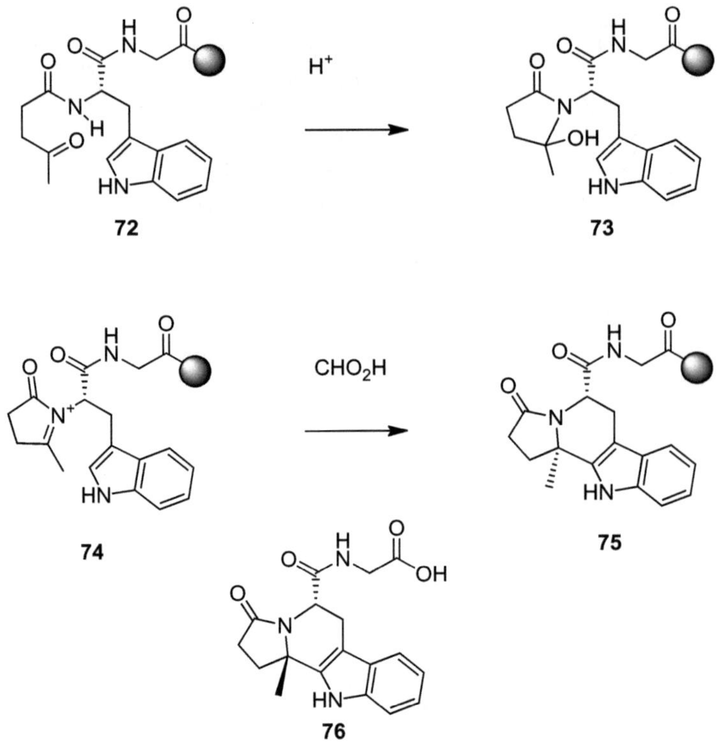

Scheme 16 Bridgehead substitution by INAIC cyclization with ketones in hot formic acid was feasible on solid support. However, the reaction is limited by significant steric bulk in the transition state

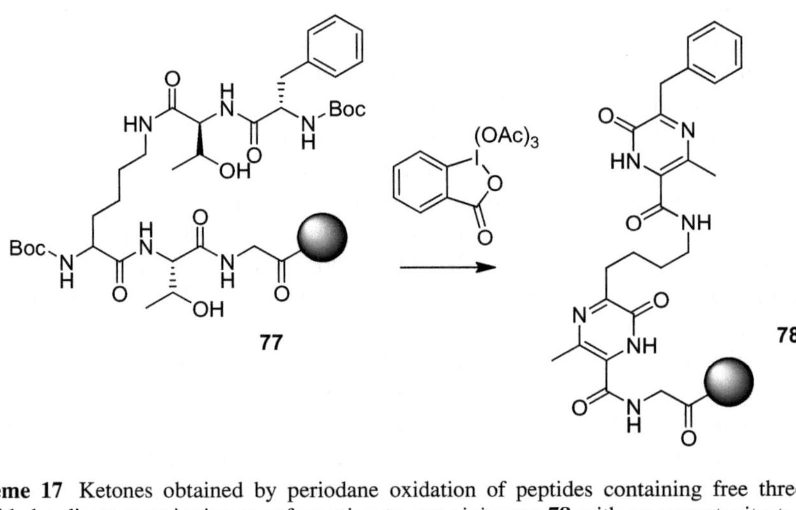

Scheme 17 Ketones obtained by periodane oxidation of peptides containing free threonines provided a direct quantitative transformation to pyrazininones **78** with no opportunity to add a second nucleophile to intermediate *N*-acyliminium ions

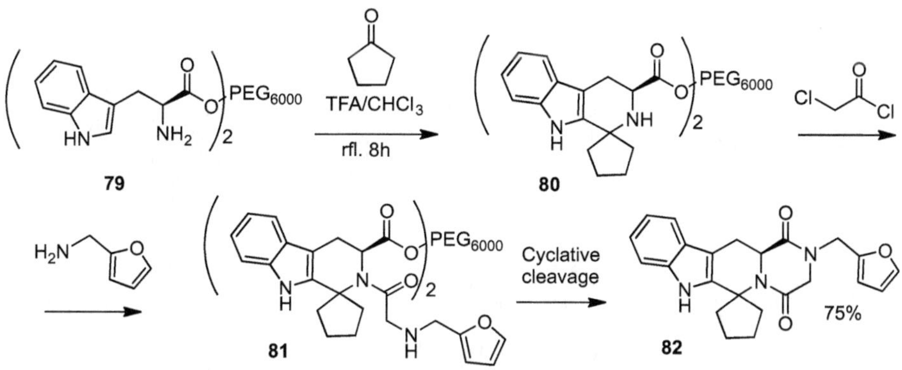

Scheme 18 The less hindered Pictet–Spengler reaction imposes less bulk than INAIC reactions in the transition state, facilitating formation of bridgehead substituted tetrahydro β-carbolines

chloroacetylation, reaction with an amine giving **81** and final cyclative release of pentacyclic **82** from the PEG support [47].

Bridgehead substitution may also be obtained by the elegant multicomponent reaction mentioned briefly above as an alternative method for preparation of *N*-acyliminium ions in solution. In this approach involving C-acylation, Michael addition, *N*-acylative annulations to give **83** and finally INAIC reaction to **84**, the *N*-acyliminium ion is formed through the annulation induced double bond migration to provide the N-acyliminium ion double bond. Thus this reaction does not involve hydroxylactam formation and provide bridgehead substitution, e.g. **85** (Scheme 19) [39].

9 Ring Closure Release Strategies

Ring closure release of peptides has been known almost as long as the technology of solid phase synthesis [48]. The generation of a nucleophile in the vicinity of an ester bond linking the target molecule to the solid support in the final step of synthesis provides a great opportunity to obtain a clean and efficient release of the desired product from the support with the benefit of generating an additional fused heterocyclic ring on the molecular scaffold. This approach has been exploited in many reports both for Pictet–Spengler and for INAIC cyclization reactions on solid support [36, 46, 47, 49–52].

Ganesans group reacted the amine of Pictet–Spengler derived tetrahydro-*β*-carboline with Fmoc amino acid chlorides and upon Fmoc deprotection the pure diketopiperazine was released [53]. In yet another approach they first let the tetrahydro-*β*-carboline nitrogen reacted with bis-p-nitrophenyl carbonate and the reactive urea formed react with an external amine which in turn reacted with the ester bond to the resin to release the fused succinimide product [50].

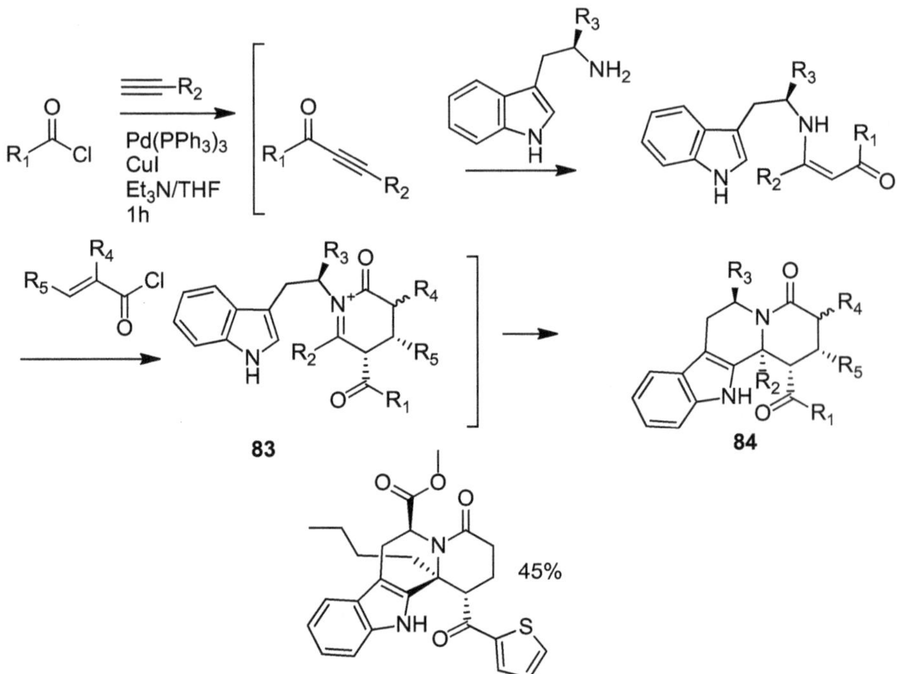

Scheme 19 Stepwise multicomponent reaction terminating with an INAIC cyclization with bridgehead substitution. The reaction has great potential for combinatorial chemistry. The reaction creates four new stereo-centres, three of which are formed under complete control of the tryptamine chirality

Scott et al. have described the release by direct reaction with the amine formed during the Pictet–Spengler condensation, see Scheme 13 [36].

Kuo et al. [52] used the methods developed by Ganesan for solid phase, in solution, reacting tryptophan methyl ester **86** with ketones and followed by cyclization of the bridgehead substituted Pictet–Spengler products **87** and yielding tetra and pentacyclic **88** and **89**, respectively. Later Chanda et al. used a variation of this strategy to release the products **82**, Scheme 18 from a soluble PEG-polymer carrier (Scheme 20).

In what was probably an attempt to reduce and reductively aminate heptanal with the azide, **90** Komnatnyy et al. observed the ring closure release of compound **91** in which the tetrahydro-β-carboline is bridged by a 6-membered lactam [46]. They could also reduce the azide to the free amine followed by hydrolysis of the resin ester in high yield. Furthermore, they coupled the azide with a variety of alkynes to give an interesting scaffold **92** with perpendicular substitution with 1,4-substituted triazoles using CuI and CuAAC conditions (Scheme 21) [1, 46].

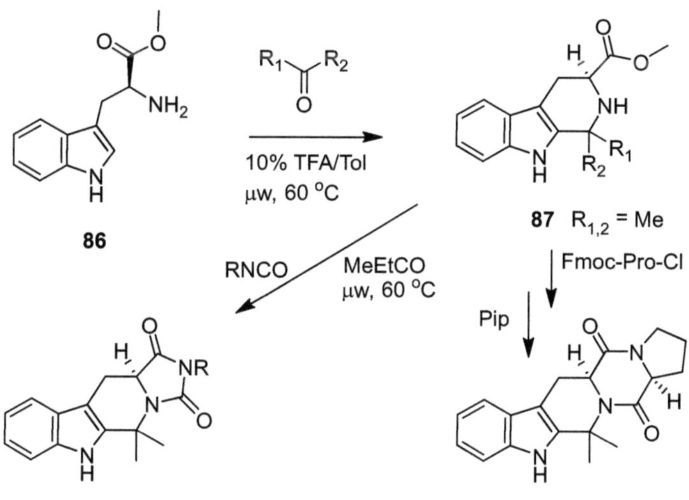

Scheme 20 In the ring closure release strategy a nucleophile is introduced in the vicinity of an ester bond, typically at the tetrahydro-β-carboline nitrogen to attack the ester with formation of an additional ring. The ester can be the linkage to a solid support or as shown here, e.g. a methyl group

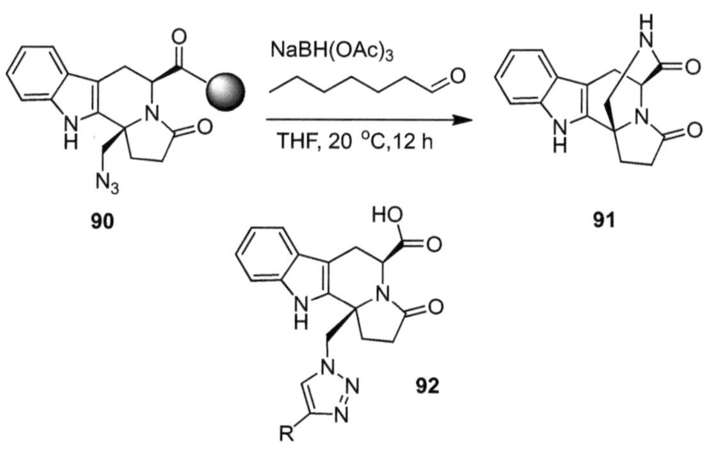

Scheme 21 The introduction of a versatile bridgehead azidomethyl group allowed synthesis molecules with interesting spatial arrangements

10 Imidazoles and Pyrroles as C-Nucleophiles

Imidazoles **93** can act as C-nucleophiles in conventional Pictet–Spengler reactions with arylaldehydes on solid support (Scheme 22). The reaction was performed at elevated temperature and neutral conditions where the imidazole was not

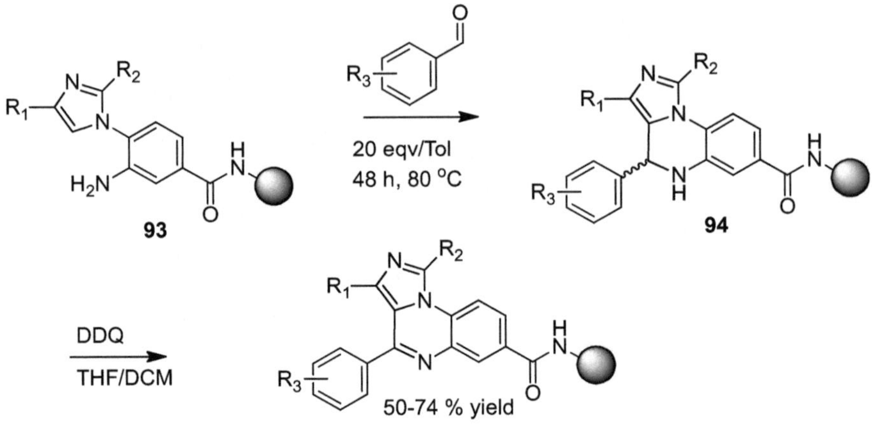

Scheme 22 The application of imidazoles as nucleophiles in solid phase Pictet–Spengler reactions using external electrophile and with oxidation of the product formed

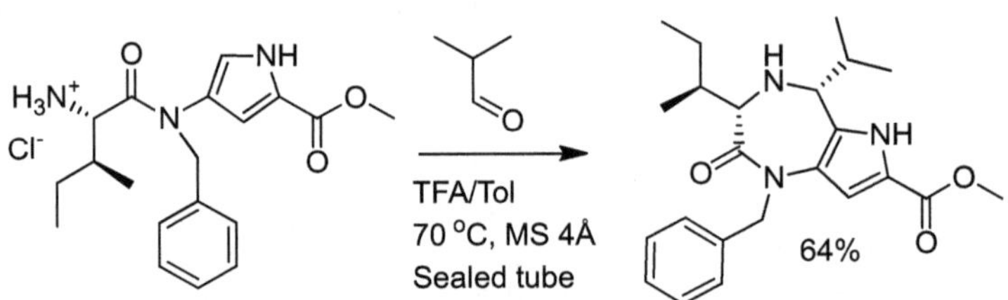

Scheme 23 Reactivity of pyrrole and cis-orientation of *N*-alkylated peptide bond allowed the 7-membered ring to be formed by Pictet–Spengler reactions during this elegant diazepine synthesis

protonated. The Pictet–Spengler product **94** was oxidized to the fully conjugated compound **95** in fair to good yields [54].

In yet another elegant approach towards diazepines the seven-membered diazepine **97** fused to a pyrrole ring was obtained by solution phase reaction of the appropriate short peptide-methyl ester **96** with aldehydes under acid catalysed Pictet–Spengler conditions. Here the barrier for reaction for the relatively unfavourable seven-membered ring and its transition state was compensated by the high nucleophilic reactivity of the two-position of the pyrrole ring and the amide alkylation, promoting the desired *cis* orientation of the central peptide bond (Scheme 23) [55].

11 Heteroatom Nucleophiles in INAIC Reactions

While C-nucleophiles have dominated the spectrum of solid phase INAIC chemistry, N-, O- and S- may also act as nucleophiles and give useful heterocyclic architectures [23]. In the course of such reactions the N-acyliminium ion reacts with an amine, alcohol or a sulfhydryl group to form cyclic *N,N*-, *N,O*- or *N,S*-acetals **100** that due to the nature of the fused cyclic products are actually relatively stable and survive most chemical manipulations (see e.g. Scheme 13). If there is no good nucleophile in the vicinity (e.g. X = CR) linked to sidechains of the peptide when the highly reactive *N*-acyliminium ion **99** is formed, this will in fact react even with a subsequent amide bond to produce the interesting scaffold **98**, Scheme 24.

In an investigation of the influence of ring-size on the outcome of the INAIC reaction it was found that the combinations for the formation of the first and second ring of 5,5- 5,6- 5,7-, 6,5 and 6,6-membered rings (Scheme 24) gave excellent yields while any combinations in which the first ring was 7-membered or the second ring was 8-memberes did not yield significant amounts of product. The combination 6,7-membered also gave a poor result. Serine and lysine did not perform well as the nucleophile, X in **99** [30].

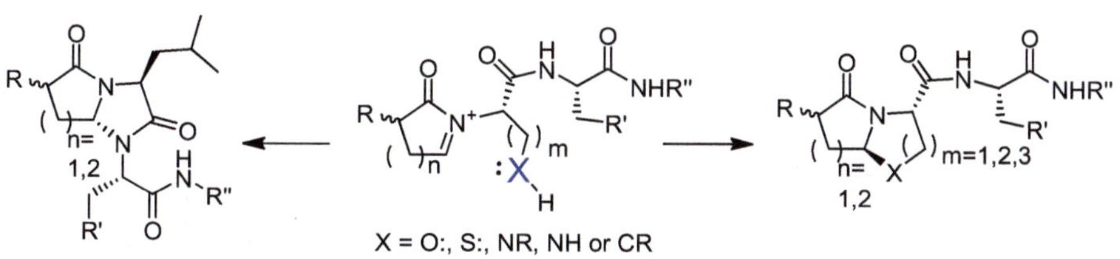

98 **99** **100**

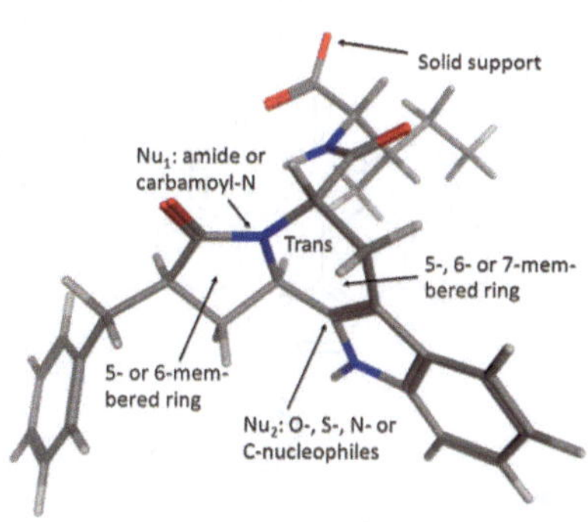

Scheme 24 An exploration of the limitations of ring-size and nucleophiles other than aromatic nucleophiles indicate many different nucleophiles can be used and the ring-size combinations 5,5- 5,6- 5,7-, 6,5 and 6,6-membered rings readily obtained

12 Extension of INAIC Products by Palladium Catalysed Cross Coupling Reactions

The rates of the INAIC cyclization reactions with aromatic C-nucleophiles are quite sensitive to the nucleophilic character of the aryl π-electrons. Therefore striking a balance that favour cyclization has to be considered when planning the synthesis of aryl-INIAC products for this purpose. Electronegative effect of substituents on the aromatic ring should be minimized. This is in contrast to the requirements for a good substrate for a subsequent Suzuki reaction. Therefore solid phase Suzuki reactions were optimized by coupling aryl boronic esters with peptidic aryl iodides prior to INAIC cyclization. These reactions were then performed quantitatively on the protected peptide aldehydes and subsequently, upon release of the aldehydes excellent yields were obtained in the INAIC reactions. The INAIC reactions with the aryl iodides prior to Suzuki reactions were less successful [56].

Alternatively, more reactive C-aryl-nucleophiles, e.g. a peptide containing thienylalanine or even the less reactive 2-bromothienylalanine, Scheme 25, **101** can react as nucleophile in INAIC cyclizations [22]. In case of the non-brominated thienyl analog the obtained INAIC cyclization product could be quantitatively and selectively brominated with bromine in acetic acid at the remaining 2-position of the thienyl ring to give **102**. The products were then employed in quantitative Suzuki reactions on solid support with a wide range of boronic esters.

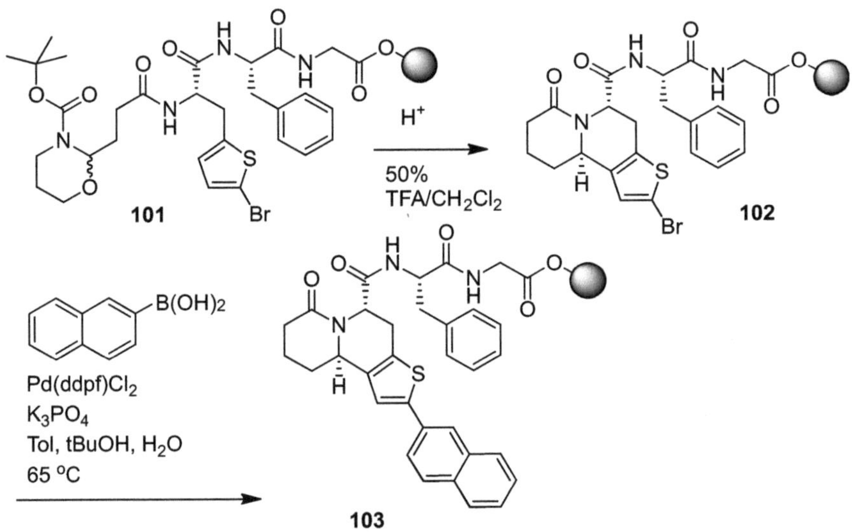

Scheme 25 Suzuki reactions can be performed pre- or post-INAIC-cyclization (on **102**) reaction. The pre-cyclization Suzuki reaction [56] (not shown) provides the best result in the subsequent INAIC-step due to increased aromatic nucleophilicity

Other types of Pd-cross coupling reactions using methylacrylate and ketene on aryl iodide were performed on solid support subsequently to Pictet–Spengler reaction with 3-iodobenzaldehyde [57].

13 Features Specific to the Solid Phase INCIC Reaction

As seen above the INAIC reactions also apply to corresponding *N*-carbamoyliminium ions [10, 26]. This approach enabled the application of amino acid derived building blocks **104** which in combination with tryptophan derivatives results in products with 1-aminomethyl tetrahydro-*β*-carboline substructure **105** [10, 25, 26]. This substructure is interestingly found in natural products, such as Woodinine [58] and Eudistomidine B [59], which maybe are formed by similar reactions in Nature (Scheme 26).

Combination of the building blocks with phenylalanine derivatives **106** led to compounds with 1-aminomethyl tetrahydroisoquinoline structure **108**. The 1-aminomethyl tetrahydroisoquinoline substructure is found in natural products such as ecteinascidin and saframycins [60]. However, as the initial *N*-carbamoyliminium ion may also rearrange reversibly to the corresponding imidazolones [26, 61] only electron rich π-nucleophiles worked for these reactions. The generated imidazolones **109** were, however, excellent substrates for Chan-Lam couplings providing **110** (Scheme 27) [26].

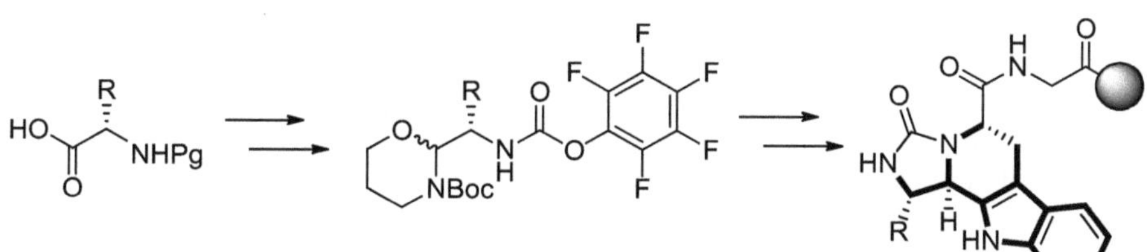

104 **105**

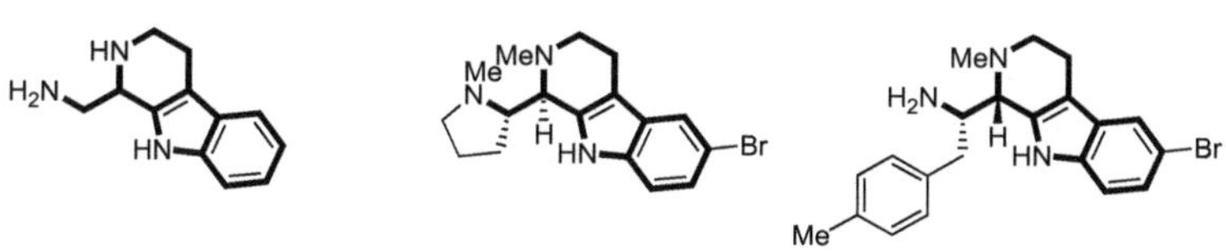

A Woodinine Eudistomidine B

Scheme 26 Synthesis of INCIC product with 1-aminomethyl tetrahydro-*β*-carboline (A) substructure which is found in Natural Products

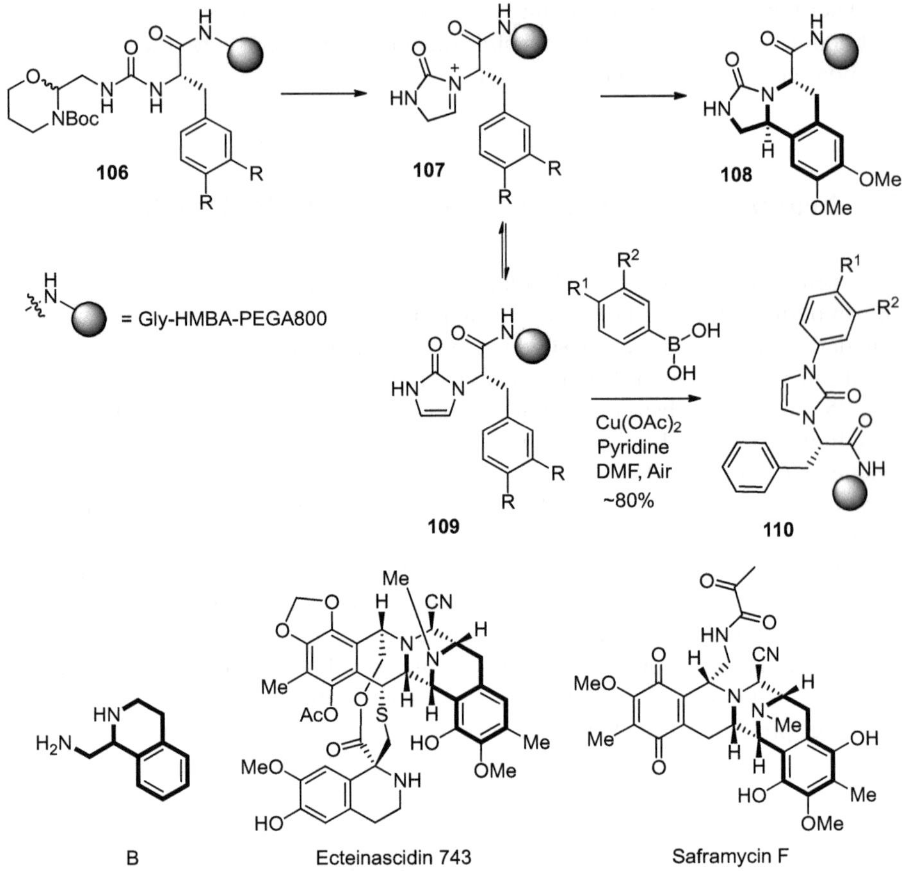

Scheme 27 INCIC reactions leading to products with 1-aminomethyl tetrahydroisoquinoline substructure (B), which is prevalent in natural products. The *N*-carbamoyliminium ion **107** could also isomerize to imidazolones, which were further functionalized through Chan-Lam couplings

Expanding the cyclic *N*-carbamoyliminium ion to a six-membered ring, **112**, circumvented the problems with the rearrangement reaction and enabled the application of also non-activated π-nucleophiles (Fig. 4).

Introduction of analogous benzaldehyde derived building blocks was also feasible. When the benzaldehyde carbamate **115** was used in combination with Trp the resulting heterocycle contained no less than five rings connected in an interesting non-planar arrangement, **117**. Compounds **117** may be fully converted by acid catalysed oxidation in presence of oxygen [10] or DDQ to stable *N*-carbamoyliminium ions **118** and **119**, which display strong fluorescence and have

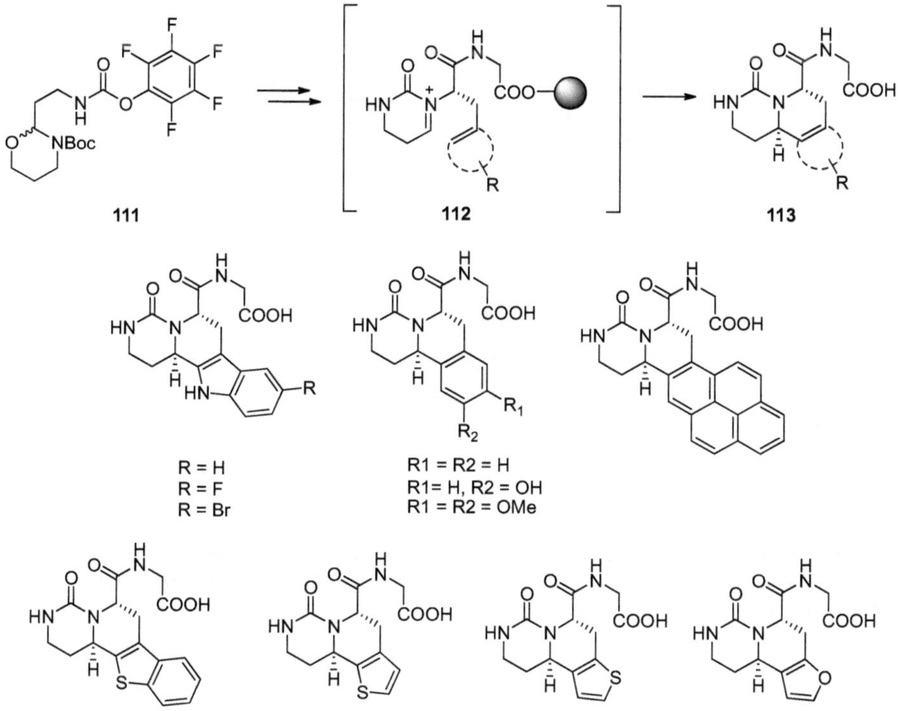

Fig. 4 Aliphatic 6,6-membered INCIC cyclization products are readily obtained with building blocks derived from 3-amino-propanal

been used for cellular labelling experiments. These novel dyes can be generated in situ in the peptide chain under oxidative conditions (Scheme 28).

Finally, it has also been demonstrated that that the N-carbamoyliminium ion derived imidazolenes are excellent as π-nucleophiles. Building blocks with imidazolones incorporated in the side chain on solid support, e.g. **120**, were reacted with Pfp carbamate **111** and formed imidazolone containing heterocycle **121** in high yield (Scheme 29).

14 Conclusion

The development of the INAIC reaction on solid support has greatly expanded the scope for the generation of molecular templates and pharmaceutically interesting scaffolds for positioning pharmacophores in various locations in 3D-space.

The richness of the chemistry is greatly enhanced by the full compatibility of the INAIC chemistry developed for peptide assembly on solid support. The

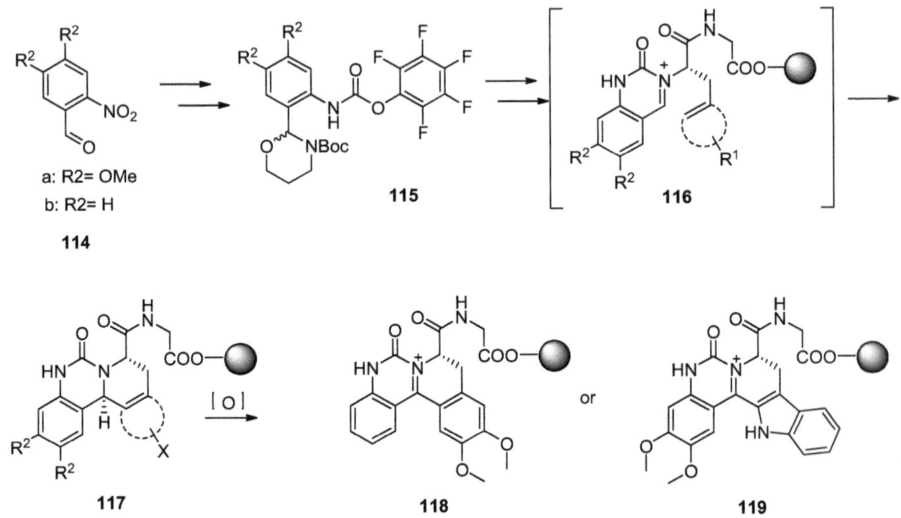

Scheme 28 INCIC products derived from 2-amino-benzaldehyd building blocks gives rise to highly fluorescent compounds by oxidation with DDQ of e.g. **117** to the cyanine like fluorophores **118** or **119**. This can be performed in situ to generate fluorescence labelled bioactive peptides

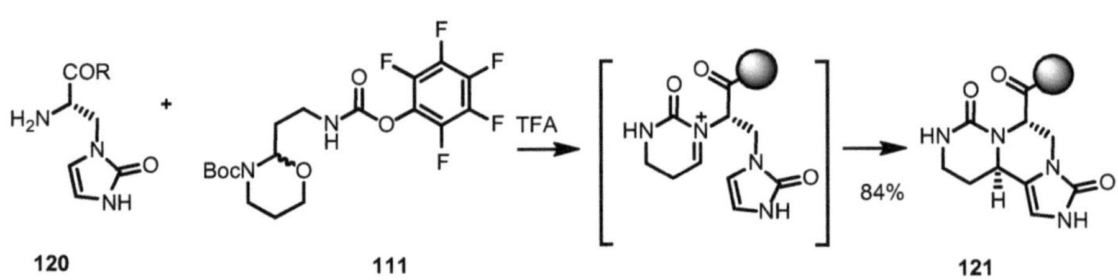

Scheme 29 Imidazolones are excellent nucleophiles for the INCIC reactions

combination of functional group accessibility in the peptide world with the complex molecular structures obtained by the INAIC reaction provides an almost ideal toolbox for preparation of combinatorial libraries of pharmaceutically interesting compounds. Some of the scaffold diversity available from solid phase INAIC chemistry is presented in Fig. 5. However, the architectures obtainable by INAIC chemistry is far from exhausted and an endless line of templated nucleophiles Nu_1 and Nu_2 (Fig. 1) and even alternative electrophiles are waiting for us to be exploited in design and screening for drugs, catalysts and new materials.

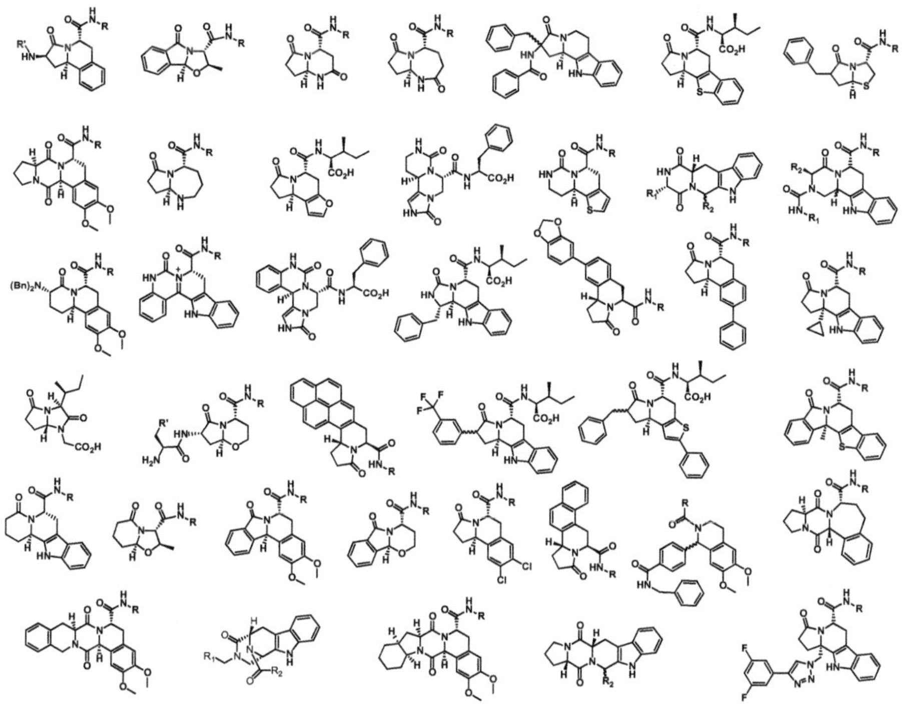

Fig. 5 The solid phase INAIC-chemistry provides a unique platform for the synthesis of different scaffold architectures that has the property characteristics similar to those of drug molecules. The chemistry furthermore takes advantage of peptide diversity and is ideally suited for synthesis and screening in a combinatorial setting

References

1. Tornøe CW, Meldal M (2001) Wave of the future. In: Lebl M, Hougten R (eds) Proceedings of 17th American peptide symposium. Kluwer Academic Publishers, San Diego, pp 263–264. doi:10.1007/978-94-010-0464-0_119
2. Tornøe CW, Christensen C, Meldal M (2002) J Org Chem 67:3057–3064
3. Rostovtsev VV, Green LG, Fokin VV, Sharpless KB (2002) Angew Chem Int Ed 41:2596–2599
4. Meldal M, Tornøe CW (2008) Chem Rev 108:2952–3015
5. Diness F, Schoffelen S, Meldal M (2015) Advances in merging triazoles with peptides and proteins, Topics in Heterocyclic Chemistry. Springer, Berlin/Heidelberg, pp 1–38. doi:10.1007/7081_2015_192
6. Mannich C, Krösche W (1912) Arch Pharm 250:647–667
7. Petasis NA, Akritopoulou I (1993) Tetrahedron Lett 34:583–586
8. Pictet A, Spengler T (1911) Chem Ber 44:2030–2036
9. Nielsen TE, Meldal M (2004) J Org Chem 69:3765–3773
10. Diness F, Beyer J, Meldal M (2006) Chem Eur J 12:8056–8066
11. Rademann J, Meldal M, Bock K (1999) Chem Eur J 5:1218–1225
12. Maryanoff BE, Zhang H-C, Cohen JH, Turchi IJ, Maryanoff CA (2004) Chem Rev 104:1431–1628

13. Cox ED, Cook JM (1995) Chem Rev 95:1797–1842
14. Brase S, Gil C, Knepper K (2002) Bioorg Med Chem 10:2415–2437
15. Meldal M, Tornoe CW, Nielsen TE, Diness F, Le Quement ST, Christensen CA, Jensen JF, Worm-Leonhard K, Groth T, Bouakaz L, Wu B, Hagel G, Keinicke L (2010) Biopolymers 94:161–182
16. Le Quement ST, Petersen R, Meldal M, Nielsen TE (2010) Biopolymers 94:242–256
17. Nielsen TE, Meldal M (2009) Curr Opin Drug Discov Devel 12:798–810
18. Groth T, Meldal M (2001) J Comb Chem 3:45–63
19. Nielsen TE, Meldal M (2005) J Comb Chem 7:599–610
20. Casnati G, Dossena A, Pochini A (1972) Tetrahedron Lett 52:5277–5280
21. Sharma A, Singh M, Rai NN, Sawant D (2013) Beilstein J Org Chem 9:1235–1242
22. Le Quement ST, Nielsen TE, Meldal M (2008) J Comb Chem 10:447–455
23. Nielsen TE, Le Quement ST, Meldal M (2005) Org Lett 7:3601–3604
24. Lee SC, Park SB (2006) J Comb Chem 8:50–57
25. Diness F, Beyer J, Meldal M (2004) QSAR Comb Sci 23:130–144
26. Diness F, Meldal M (2009) Chem Eur J 15:7044–7047
27. Nahm S, Weinreb SM (1981) Tetrahedron Lett 22:3815–3818
28. Wen JJ, Crews CM (1998) Tetrahedron 9:1855–1858
29. Christensen C, Tornøe CW, Meldal M (2004) QSAR Comb Sci 23:109–116
30. Le Quement ST, Nielsen TE, Meldal M (2007) J Comb Chem 9:1060–1072
31. Berry JM, Howard PW, Thurston DE (2000) Tetrahedron Lett 41:6171–6174
32. Dahlqvist A, Olsson I, Nordén Å (1965) J Histochem Cytochem 13:423–430
33. Geoghegan KF, Stroh JG (1992) Bioconjug Chem 3:138–146
34. Li X, Zhang L, Hall SE, Tam JP (2000) Tetrahedron Lett 41:4069–4073
35. Petersen R, Le Quement ST, Nielsen TE (2014) Angew Chem Int Ed 53:11778–11782
36. Scott WL, Martynow JG, Huffman JC, O'Donnell MJ (2007) J Am Chem Soc 129:7077–7088
37. Nielsen TE, Meldal M (2005) Org Lett 7:2695–2698
38. Groth T, Meldal M (2001) J Comb Chem 3:34–44
39. Karpov AS, Oeser T, Müller TJJ (2004) Chem Commun 1502–1503
40. Hansen CL, Clausen JW, Ohm RG, Ascic E, Le Quement ST, Tanner D, Nielsen TE (2013) J Org Chem 78:12545–12565
41. Ascic E, Jensen JF, Nielsen TE (2011) Angew Chem Int Ed 50:5188–5191
42. Silveira CC, Felix LA, Braga AL, Kaufman TS (2005) Org Lett 7:3701–3704
43. Diness F, Meldal M (2015) Eur J Org Chem 7:1433–1436
44. Lee SC, Choi SY, Chung YK, Park SB (2006) Tetrahedron Lett 47:6843–6847
45. Li X, Zhang L, Zhang W, Hall SE, Tam JP (2000) Org Lett 2:3075–3078
46. Komnatnyy VV, Givskov M, Nielsen TE (2012) Chem Eur J 18:16793–16800
47. Chanda K, Chou CT, Lai JJ, Lin SF, Yellol GS, Sun CM (2011) Mol Divers 15:569–581
48. Fridkin M, atchornik A, Katchalski E (1965) J Am Chem Soc 87:4646–4648
49. Ganesan A (2006) Mini Rev Med Chem 6:3–10
50. Bonnet D, Ganesan A (2002) J Comb Chem 4:546–548
51. Fantauzzi PP, Yager KM (1998) Tetrahedron Lett 39:1291–1294
52. Kuo FM, Tseng MC, Yen YH, Chu YH (2004) Tetrahedron 60:12075–12084
53. Wang H, Ganesan A (1999) Org Lett 1:1647–1649
54. Kundu B, Sawant D, Chhabra R (2005) J Comb Chem 7:317–321
55. Deaudelin P, Lubell WD (2008) Org Lett 10:2841–2844
56. Nielsen TE, Le Quement S, Meldal M (2005) Tetrahedron Lett 46:7959–7962
57. Dondas HA, Grigg R, MacLachlan WS, MacPherson DT, Markandu J, Sridharan V, Suganthan S (2000) Tetrahedron Lett 41:967–970
58. Rinehart KL, Kobayashi J, Harbour GC, Hughes RG Jr, Mizsak SA, Scahill TA (1984) J Am Chem Soc 106:1524–1526
59. Kobayashi J, Cheng JF, Ohta T, Nozoe S, Ohizumi Y, Sasaki T (1990) J Org Chem 55:3666–3670
60. Scott JD, Williams RM (2002) Chem Rev 102:1669–1730
61. Diness F, Meldal M (2010) Pept Sci 94:236–241

Top Heterocycl Chem (2017) 52: 95–138
DOI: 10.1007/7081_2016_4

Published online: 26 March 2017

Advances in Metal-Mediated Solid-Phase Synthesis of Heterocyclic Compounds

Agustina La-Venia, Carina M.L. Delpiccolo, and Ernesto G. Mata

Abstract While heterocycles are a key feature frequently found in compounds with pharmaceutical or agrochemical interest, the combination between solid-supported synthesis and organometallic chemistry has demonstrated to be a helpful tool for the preparation of that class of biologically relevant structures. From the more disclosed Hüisgen 1,3-dipolar cycloaddition, the palladium-catalyzed synthesis of indoles and the ruthenium carbene-mediated preparation of heterocycles by ring-closing metathesis (RCM), to the less developed use of gold and iron in polymer-supported chemistry, this survey offers an overview of the solid-phase application of metal-mediated chemistry to the generation of libraries of a variety of heterocyclic systems.

Keywords Heterocycles • Organometallic chemistry • Solid-phase organic synthesis

Contents

A. La-Venia, C.M.L. Delpiccolo, and E.G. Mata (✉)
Facultad de Ciencias Bioquímicas y Farmacéuticas, Instituto de Química Rosario (CONICET-UNR), Universidad Nacional de Rosario, Suipacha 531, 2000 Rosario, Argentina
e-mail: mata@iquir-conicet.gov.ar

1 Introduction

Studies related to the synthesis of small "drug-like" molecules remain an essential strategy in identifying lead compounds for the treatment of diseases, contributing to the understanding of the molecular basis of biological mechanisms [1, 2]. Therefore, the major challenge of organic synthesis and medicinal chemistry has in recent years been the development of new processes to generate diversity and complexity in an efficient, economic, and environmentally benign way.

In the search for molecular diversity, the use of metal-mediated reactions has acquired increasing popularity [3] since they allow the formation of carbon–carbon and carbon–heteroatom (N, O, S) bonds. They are critical for obtaining complex structures of biological interest, both those related to natural products and those arising from synthetic protocols such as diversity-oriented synthesis (DOS) [4–6].

While solid-phase organic synthesis (SPOS) has evolved in the last two decades to become an interesting tool for the preparation of a large number of structurally diverse compounds for combinatorial libraries, its combination with organometallic chemistry has added new comparative advantages, including an increase in chemo- and regioselectivity [7]. Particularly, metal-mediated cross-coupling reactions have been benefited by the application of solid-supported chemistry. Excess of the non-immobilized substrate can be added to complete the reaction since the corresponding homodimer remains in solution and can be removed by filtration. On the other hand, homodimerization of the immobilized substrate is less favorable due to the separation between the reactive sites (Scheme 1). Last but not least, solid-phase synthesis makes a substantial contribution to green chemistry by reducing solvent waste comparing to traditional organic synthesis that usually requires many chromatographic separations [8].

Since heterocyclic compounds are key structures frequently found in natural products and synthetic molecules, either for pharmaceutical or agrochemical use [9], we decided in this chapter to focus on the previously unreviewed application of

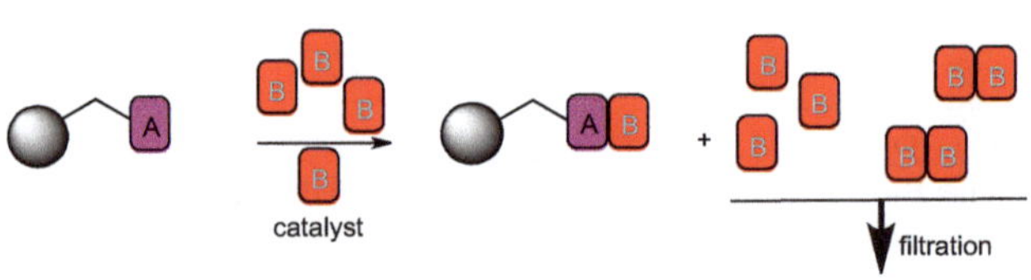

Scheme 1 Cross-coupling reaction on solid-phase synthesis

metal-mediated solid-phase chemistry to the synthesis of heterocyclic systems. For a better understanding of the compiled information, the chapter has been organized according to the metal involved in the heterocycle generation.

2 Heterocyclizations Defined by Metal

2.1 Cobalt

Cobalt complexes have gained popularity as an interesting, economical, and environmentally friendly alternative to replace, in some cases, expensive palladium and toxic nickel catalysis [10, 11]. Especially useful for solid-phase chemistry has been the application of catalytic organocobalt reagents to the generation of heteroarenes by [2 + 2 + 2] cycloisomerization. When [2 + 2 + 2] cycloisomerization is carried out in homogeneous phase, competing linear dimerization or cyclotrimerization of the substituted acetylenes can be a problem, obtaining low chemo- and regioselectivities. A solution to this can be to have one of the substrates immobilized to a resin. Particularly, in order to improve chemoselectivity in the cobalt-catalyzed three-component synthesis of pyridine moieties, one of the alkyne component was linked to a solid support, while the other alkyne and the nitrile remained in solution (Scheme 2) [12]. Thus, 2-propyn-1-ol tethered to trityl resin **1** reacted with six different alkynes **2** and three different nitriles **3** in the presence of dicarbonylcyclopentadienyl cobalt(I) [$CpCo(CO)_2$] (20 mol%) and tetramethylammonium oxide (TMAO) as a catalyst activating additive, to give the immobilized pyridines **4**. Nitriles were added in very large excess to avoid formation of the immobilized benzene derivative. Finally, TFA cleavage afforded the corresponding pyridines **5** with complete chemoselectivity, although in complex mixtures of regioisomers.

One step forward was the solid-supported, microwave-mediated [2 + 2 + 2] cycloaddition to give bicyclic pyridines, pyridones, and iminopyridines under $CpCo(CO)_2$ catalysis [13]. In this case, the strategy was based on the immobilization of diyne to trityl resin (**6**) and the reaction with excess of soluble nitriles **7** to afford, after cleavage with TFA, the corresponding dihydropyrrolopyridines **9** in

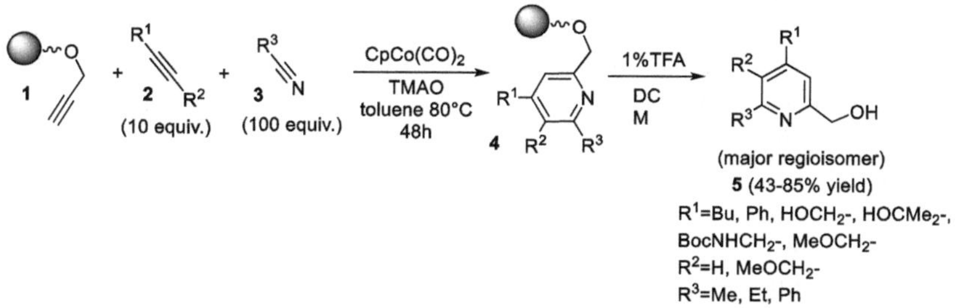

Scheme 2 Catalytic organocobalt reagents for the synthesis of pyridines by [2 + 2 + 2] cycloisomerization

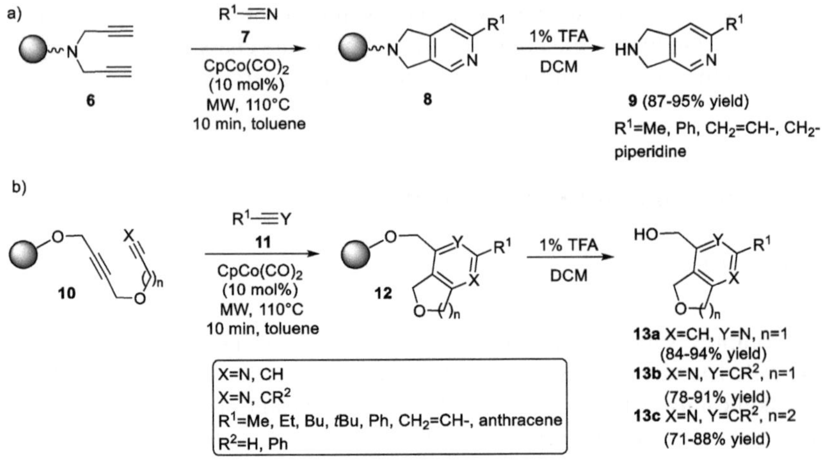

Scheme 3 Cobalt-catalyzed [2 + 2 + 2] cycloisomerization for the synthesis of bicyclic pyridines

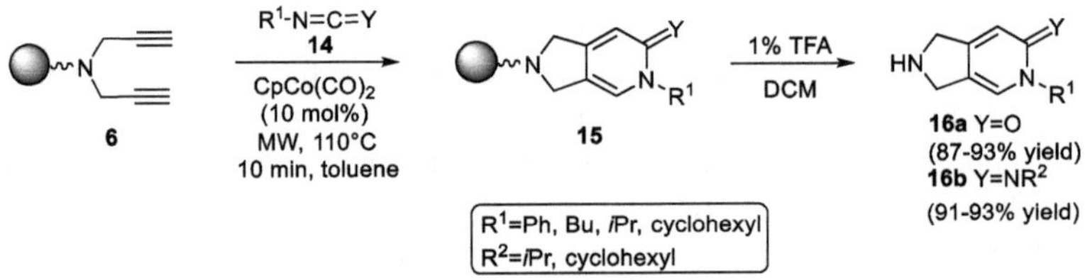

Scheme 4 Synthesis of pyridones and iminopyridines via co-promoted cyclotrimerization

very high yields (Scheme 3a). This solid-phase [2 + 2 + 2] cyclotrimerization has a clear comparative advantage over its solution-phase counterpart by avoiding side products formed due to a competitive cycloaddition between two molecules of the soluble diyne. In fact, a solution-phase version gave only a 46% yield of the pyridine with several by-products like benzene derivatives coming from homocoupling between two or three diyne molecules. Interestingly, nonsymmetrical immobilized diynes like **10** (X = CH, $n = 1$) reacted with nitriles **11** (Y = N) to finally achieve the furopyridines **13a** in high yields with complete regioselectivity (Scheme 3b). Similarly, using trityl resin-tethered alkynyl nitriles **10** (X = N, $n = 1, 2$) and soluble alkynes **11** (Y = CR^2), positional isomeric furopyridines **13b** and **13c** were obtained after separation from the resin.

When the symmetrical diyne **6** was treated with isocyanates **14** (Y = O) and carbodiimides **14** (Y = NR^2), cyclotrimerization in the presence of the same catalyst afforded the bicyclic pyridones **16a** and iminopyridines **16b** in good yields, after cleavage from the resin (Scheme 4). It is interesting to note that when the cyclotrimerization was performed under the same conditions but using conventional heating for 24 h, the expected products were not obtained.

2.2 Copper

Copper is a very versatile metal, whose catalyst properties have been known for more than one century since the finding of Ullmann reaction [14]. Even though this catalyst remained neglected for an expanded period of time, a remarkable growth in the copper-catalyzed transformations occurred over the last decades. The increasing interest in this metal must be related to many of its benefits. At first place, copper chemistry is incredibly diverse. The efficient coordination of Cu with heteroatoms and π-bonds allows an expanded scope of substrates for the Cu catalysis, taking part in many cross-coupling reactions, and terminal alkyne activation, among others. In addition, depending on the metal oxidation state, copper can participate in both one- and two-electron mechanisms (radical and polar). Secondly, copper is an abundant metal on earth, which makes its costs much lower than precious transition metals, and allows sustainable chemical transformation procedures. Moreover, the salts of copper usually present low toxicity.

During this growth of the copper catalysis application, its employment in solid-phase synthesis has been emerged as well, although mostly applied to click chemistry. The first report in which the copper catalysis was used for the generation of heterocycles on solid support was contributed by Tietze et al. in 2001 [15]. In this example, Wang resin-supported 1,2-diaza-1,3-butadienes (**17**) and β-ketoamides (**18**) in the presence of catalytic amount of $CuCl_2$ led to 4-aminocarbonylpyrroles (**19**), which were cleaved from the polymer by NaOMe/MeOH to provide the final heterocycles with acceptable to good yields (Scheme 5). The *tandem* reaction, Michael-type addition followed by an intramolecular addition of the nitrogen (N = C) to the γ-carbonyl group and final dehydration, proceeded smoothly under copper (II) ion catalysis.

In the following years, an increasing number of reports combining cooper salts and SPOS appeared, including the synthesis of fused bicyclic γ-lactams **20** (Scheme 6a) [16] and indole skeletons **21** (Scheme 6b) [17, 18]. Both cyclization conditions required the use of stoichiometric amount of Cu salts (1–2 equiv.), and MW irradiation at elevated temperature. Nevertheless, beyond doubt, the main and most productive application of Cu chemistry to solid support synthesis is established on the click chemistry towards the generation of triazole rings. The section below is dedicated to describe this chemistry in more detail.

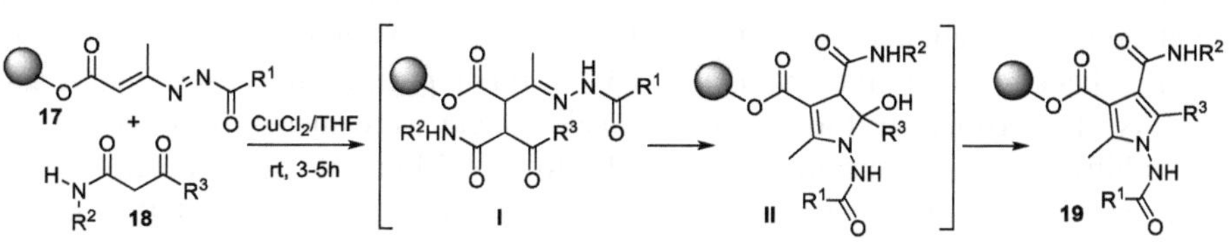

Scheme 5 Synthesis of immobilized aminocarbonylpyrroles **19** catalyzed by Cu(II)

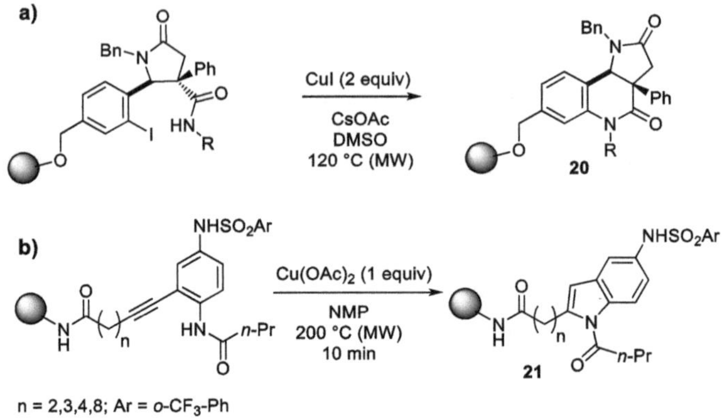

Scheme 6 Heterocyclizations on solid phase mediated by Cu(I) and Cu(II)

2.2.1 Copper-Mediated Solid-Phase Click Chemistry

Click chemistry is a modular approach that uses only the most practical and reliable chemical transformations [19]. Click chemistry has emerged as a concept that encompasses an assemblage of chemical reactions with common features: modular, wide scope, associated with very high yields, with no- or only inoffensive by-products generation, and proceeding with stereospecificity [20]. As well, the chemistry process involved must present the following characteristics: simple reaction conditions, readily available starting materials and reagents, no solvent used, or benign solvent or easily removable and simple product isolation. Due to all these advantages, the application of click chemistry has increased in the drug discovery process at different stages. In particular, the generation of 1,2,3-triazoles from azides and acetylenes, catalyzed by copper (I), has arisen as one of the favorite and most efficient application of this chemistry concept. The success of this Hüisgen 1,3-dipolar cycloaddition is based on its complete specificity and its broad biocompatibility with many reactants and functional groups, in addition to the biological importance of the triazole products themselves. Alike, the Cu(I)-catalyzed azide-alkyne cycloaddition on solid phase (CuAAC-SP) has played an essential role in the progress of supported synthesis oriented towards the development of new solid-phase strategies as well as the production of combinatorial libraries [21]. In 2001, Meldal et al. reported the first procedure for 1,3-dipolar cycloadditions mediated by Cu(I) salts to obtain 1,4-substituted [1,2,3]-triazoles in peptide backbones or side chains [22, 23]. In this report, primary, secondary, and tertiary alkyl azides, aryl azides, and an azido sugar were successfully coupled to immobilized peptidic terminal alkynes (**22**) to give **23** in mild and efficient copper catalysis conditions (Scheme 7). The reaction conditions involved the use of 2 equivalents of CuI in the presence of base at room temperature, with several solvents proved to be useful (DCM, DMF, toluene, THF, and AcCN).

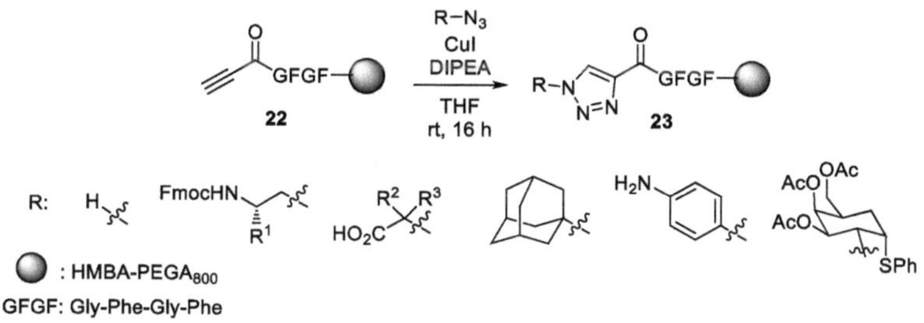

Scheme 7 Copper(I) mediated cycloadditions to yield peptide triazoles

Since then, the CuAAC-SP has expanded outstandingly, addressing a ubiquitous range of applications: modification of peptides, nucleotides, small molecules, supramolecular structures, and polymers. The main variables to define in this kind of reactions are the polymeric support, the Cu source, solvent, and presence or absence of base. As it is illustrated with the examples below, there is a wide assortment of cycloaddition conditions and an enormous area of the chemical space that can be explored, denoting the remarkable potential of this chemistry. The condition combinations may include different resins such as Merrifield, Wang, Rink amide, TentaGel, PEGA800, and AMCM, in addition to several cationic copper sources: CuBr, CuI, CuCl, $CuBr(PPh_3)_3$ (widely used with a Cu (I) stabilizer), and various polar solvents (polarity of the solvents is crucial to achieve high yields). Although not imperative, a base is usually used to improve performance. In general, both functionalities, azide and alkyne, can be either in solution or attached to the polymer. Recently, a very exhaustive review describing general features and the broad area of applications of CuAAC-SP has been published by Albericio et al. [21].

Click chemistry as source of triazole heterocycles has been combined with the derivatization of peptides in very versatile manners [24]. The cycloaddition between supported alkyne-functionalized peptide with a variety of organic azides led to conjugated peptides allowing, in some cases, the prolongation of the peptidic backbone after the triazole formation (Scheme 7) [22, 23]. The scope of this strategy allowed the synthesis of a one-bead-two-compound library of peptidotriazoles, which exhibited biological effect against a recombinant cysteine protease, *Leishmania mexicana* [25]. Similarly, the solid-phase conjugation of the antimicrobial cyclic decapeptide BPC194 to a linear or cyclic sequence through a 1,2,3-triazole ring was achieved by CuI in the presence of ascorbic acid [26]. In fact, there are an increasing number of examples of interesting couplings of triazole to side-chain sites of peptidic structures in order to modulate their properties [27, 28], as well as to achieve peptidic fragments ligation at particular position of their side chains [29]. Furthermore, click chemistry has been explored to overcome the difficulties associated with the generation of cyclic peptides due to the formation of dimers or even oligomers during the cyclization step. For instance, the solid-

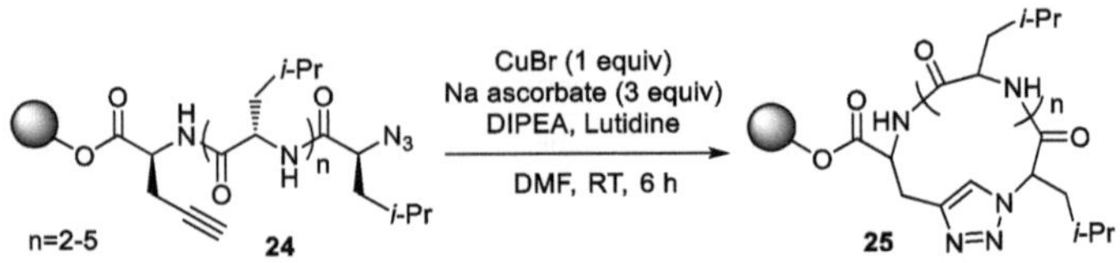

Scheme 8 Click macrocyclization of oligopeptides

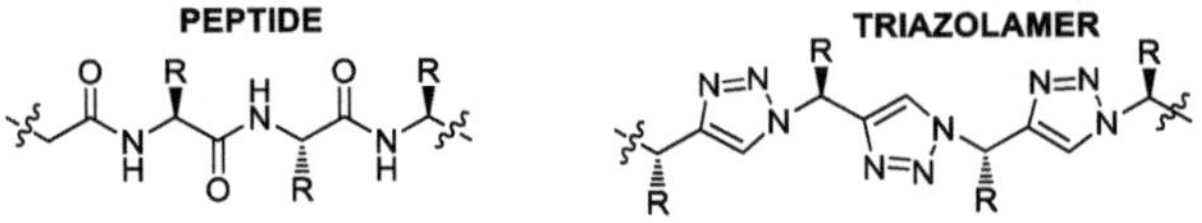

Fig. 1 Triazole-based biomimetics oligomers

phase dipolar cycloaddition catalyzed by Cu(I) facilitated the macrocyclization of cyclic tetra-, penta-, hexa-, and heptapeptides **24** where a peptide bond has been replaced by the triazole in the skeleton of the final macrocycle **25** (Scheme 8) [30]. Similarly, peptide cyclization and head-to-tail cyclodimerizations of resin-bound oligopeptides bearing azide and alkyne groups were achieved by Cu (I) catalysis [31, 32].

Actually, the replacement of peptide bonds by triazole rings made accessible a new class of non-peptidic compounds able to display protein-like functionality: the triazolamers, which may effectively mimic a β-strand (Fig. 1). These new class of oligomers that contain amino acids residues linked by 1,2,3-triazoles were described by Arora et al. in 2005 [33]. The synthetic strategy for these triazole-based biomimetics oligomers involves precisely an iterative sequence consisting of diazotransfer and Hüisgen 1,3-dipolar cycloaddition steps. Briefly after, this group described the efficient synthesis of triazolamers (5 and 6 residues) on solid support (PAM resin); however, on the contrary to their optimized conditions in classical solution phase, Zn(II) showed to be more efficient diazotransfer agent than Cu (II) (reduced in situ) [34]. Apart from the examples above [30, 34], this concept has been successfully applied to the synthesis of three peptidomimetics, analogues of Leu-enkephalin, in which the position of the bond-mimetic heterocycle in the peptide backbone was systematically changed [35]. Similarly, small peptidomimetics of the Grb2-SH2 domain with potential anticancer activity (**27**) were synthesized on solid support (Wang resin) from azide-derivatized peptides **26** via the key Cu(I)/His-mediated self-activating Hüisgen [3 + 2] cycloaddition (Scheme 9) [36].

The synthesis of peptide conjugates has taken advantage of the extremely high selectivity of the azide-alkyne cycloaddition and its compatibility with an extensive variability of functional groups. In this manner, a set of glycosylated peptoids **29** have been prepared via a click reaction by conjugation between glycosyl residues and immobilized peptoids **28** (Scheme 10) [37]. Concerning peptide nucleic acids (PNAs) [38], the Cu(I)-catalyzed [2 + 3] azide/alkyne cycloaddition has been applied as key reaction for conjugation of immobilized PNAs with different

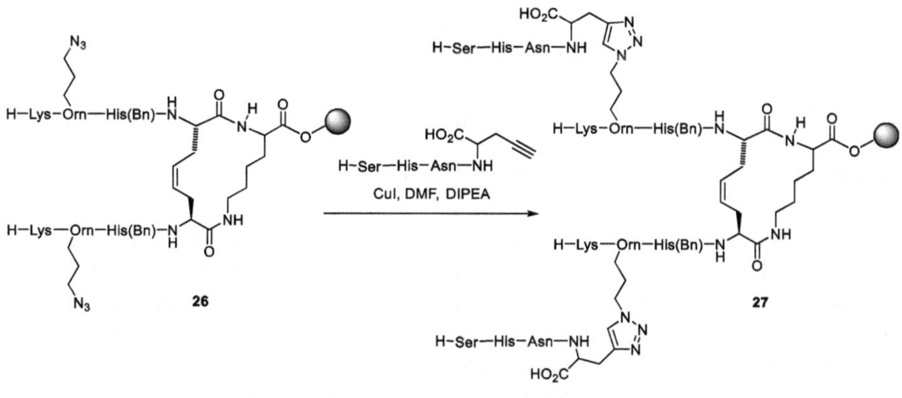

Scheme 9 Synthesis of conformationally fixed cyclic peptidomimetic **27** with 1,2,3-triazole rings incorporated in the skeleton

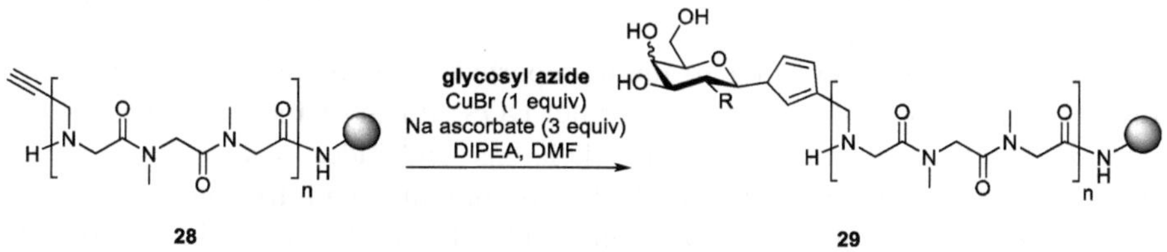

Scheme 10 Conjugation between glycosyl azide residues and immobilized peptoids **28**

labeling molecules: I) organometallics compounds – (a) ferrocene (PEGA800 resin) [39], (b) azidoferrocene, ethynylferrocene, and DEPA-ferrocene derivatives (TentaGel resin) [40], (c) azidomethyl-ruthenocene (TentaGel resin) [41], and (d) Dap (2,2′-dipicolylamine) (TentaGel resin) [42] and II) fluorescent moieties – (a) rhenium tricarbonyl complex of a bis(quinoline)-derived ligand (TentaGel resin) [43] and (b) polycyclic aromatic hydrocarbon ligand (TentaGel resin with Rink amide linker) [44]. In addition, peptide–RNA covalent conjugates have also been prepared using click conjugation [45].

In addition to click chemistry related to peptides previously described, CuAAC strategy has provided the simple method for the modification of nucleotide-containing structures, both in solution chemistry and on solid support [46–48]. CuAAC-SP has been extensively applied to diverse final targets, such as nucleoside bioconjugates [47], nucleoside derivatives libraries [49], glyconucleotides [50, 51], fluorescent DNA probes [52], just to mention some.

In a similar manner, the triazoles induced by Cu have been associated with several privileged heterocyclic structures, as it is disclosed in the following examples. For instance, valuable triazolyl aminoacyl(peptidyl) penicillins were produced using a versatile and efficient methodology on solid-phase whose crucial step consisted in a Cu(I)-catalyzed Hüisgen 1,3-dipolar cycloaddition [53]. The design of the final conjugated compounds assured the inclusion of two heterocyclic moieties (β-lactam and triazol) with numerous potential biological activities and a

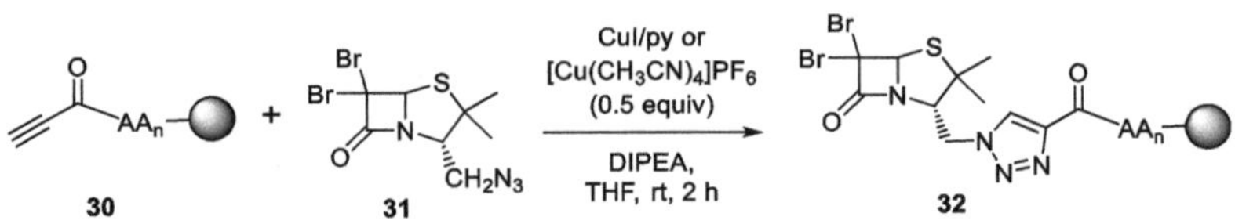

Scheme 11 Synthesis of immobilized triazolyl aminoacyl (peptidyl) penicillins **32**

peptidic fragment that may be related to cell membrane transport and recognition processes. The synthetic sequence started with traditional peptide solid-phase synthesis on Wang resin, followed by final alkynyl functionalization (**30**) to allow the coupling of azide-derivatized β-lactam (**31**) via the key cycloaddition (Scheme 11). The optimized conjugation conditions involved a source of Cu(I) (CuI/pyridine or organic soluble [$Cu(CH_3CN)_4$]PF_6) in the presence of DIPEA. This library of triazolyl aminoacyl(peptidyl) penicillins (**32**) was evaluated for their antiproliferative activity against HeLa and B16-F0 cell lines, resulting, in some cases, in high anticancer activity simultaneously with a promising selectivity profile against normal cells [54].

In addition, an interesting example reported by Park et al. consisted in a divergent and practical solid-phase parallel DOS strategy which was successfully applied for the generation of core skeletons embedded with the privileged benzopyranyl substructure, one of which was combined with triazole heterocycles via CuAAC [55]. Similarly, but at larger scale, during the synthesis of a 10,000-membered library of molecules resembling the natural product carpanona, triazole coupling catalyzed by Cu(I) was utilized for beneficial derivatizations [56].

The benefits of click chemistry have also been exploited for the development of linkers for solid support with different properties, which allow orthogonal protection and milder final cleavage conditions maintaining, or even improving, both global yields and final compounds purity [57, 58]. In particular, Gmeiner and coworkers have taken profitable advantage of click chemistry to produce an efficient and high-yielding process for the immobilization of the functional linker unit via the triazole moiety. Based on this concept, versatile functionalized resins were developed, introducing novel families of resins: (a) BAL: backbone amide linkers [59], (b) REM: regenerative Michael acceptors [60], and (c) scavenger resins [61]. Some specific polymers of these families are portrayed in Fig. 2: triazolylmethyl acrylate resin (TMA resin) (Fig. 2a), substituted phenylether-linked resins (Fig. 2b), formylindolylmethyltriazole resin (FIMT resin) (Fig. 2c), and formylpyrrolylmethyltriazole resin (FPMT resin) (Fig. 2d) [58–64].

Although related details are beyond the purpose of this chapter, it is worthy to mention that the CuAAC on solid support strategy has and still is actively employed for the creation of new materials, such as glycodendrimers [65], dendritic cores containing star polymers [66], and HPLC packings [67], among others.

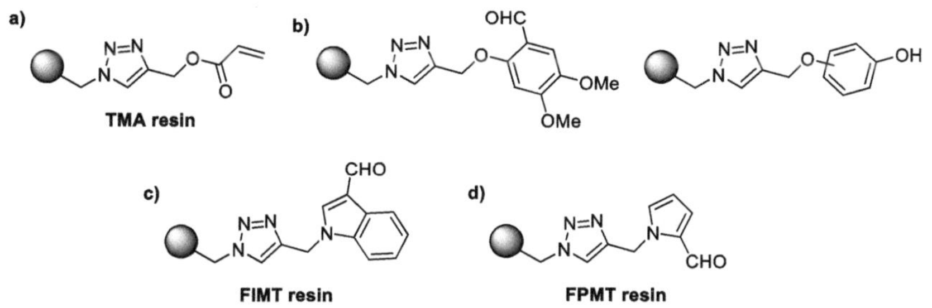

Fig. 2 Click-triazole resins developed by Gmeiner's group

2.3 Gold

Over recent years, research on homogeneous gold catalysis has reached a remarkable high level of development due to its ability of enabling a series of challenging transformations that lead to new carbon–carbon or carbon–heteroatom bonds [68–76]. Gold is a "soft" transition metal that presents a strong alkynophilicity, which is evidenced by the high binding energies of alkyne–gold complexes, in addition to fully relativistic computational analysis of gold's superiority regarding this kind of activation in comparison to different metals [77, 78]. Therefore, gold is able to activate carbon–carbon triple bonds efficiently, making them susceptible to nucleophilic attack by heteroatoms such as N, O, and S [79]. This enhanced ability for the π-activation not only defines gold reactivity but also a good functional group compatibility [80]. On the other hand, the unusual low oxophilicity presented by gold, contrary to most Lewis acids, provides to the auric catalysis an extraordinary advantage, as oxygen, water, and alcohols can be tolerated, enabling mild reaction conditions [81]. While the outstanding advances of this metal catalyst in solution chemistry are in continuous evolution, heretofore the expansion of auric-promoted transformations on polymer supports has remained scarce. In the last years, two reports have been published proving the efficiency of the auric-promoted cyclizations on solid support and promoting further application of gold chemistry on solid-phase synthesis.

The first solid-phase example of gold-promoted transformation has been reported in 2012. A gold-catalyzed heterocyclization of immobilized 2-alkynylanilines (**33**) was developed as the key step in the synthetic sequence leading to a set of 2-substituted indoles **35** (Scheme 12) [82]. In this report, several gold catalysts, such as AuCl, $AuCl_3$, and $AuClPPh_3$, have been screening, to conclude that the optimized conditions consisted in the use of 5–10 mol% of simple salt AuCl, in DCM for 4 hours at room temperature. Thus, a small library of indoles **35a–f** was generated with acceptable global yields, and high yields for the key auric-mediated cyclization step, especially for aromatic substitution at R position (over 93%). The authors emphasized that the success of the supported heteroannulation towards heterocycles **34** strongly depended on an efficient work

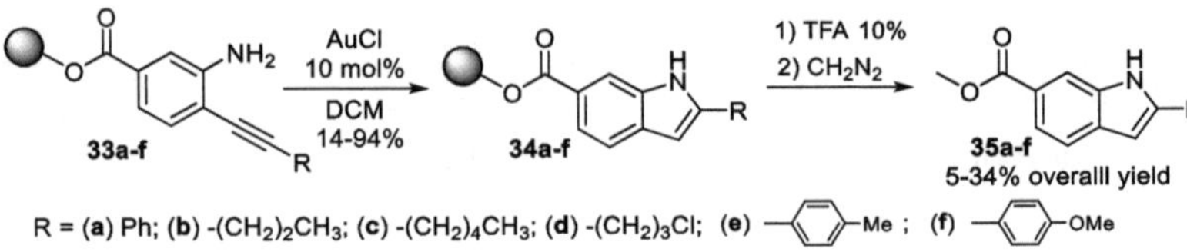

Scheme 12 Gold(I)-catalyzed generation of 2-substituted indoles (**35**)

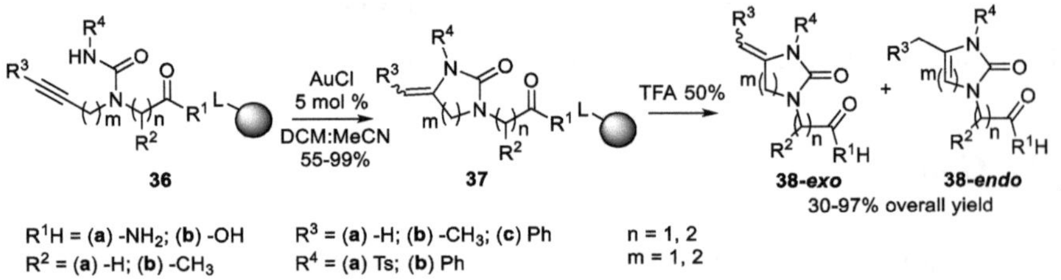

Scheme 13 Synthesis of substituted imidazolidin-2-ones (**38-*exo***) and imidazol-2-ones (**38-*endo***) via gold-promoted cycloisomerization

up of the tin-mediated nitroreduction, a previous step in the sequence leading to lineal substrates **33**.

A second example regarding gold-mediated cyclizations on solid phase has recently been published, enlarging the applicability of this metal catalyst for immobilized transformation. In this case, small library of imidazolidin-2-ones (**38-*exo***) and imidazol-2-ones (**38-*endo***) was carried out employing a high chemo- and regioselective gold-catalyzed cycloisomerization of supported propargylureas (**36**) as a key step (Scheme 13) [83]. The final heterocycles were obtained with good to excellent total yields; moreover, the key metal-promoted annulation proceeded over 55%. A full catalyst system screening was carried out, including diverse Au salts and complexes combined with silver cocatalysts in different reaction conditions, for both immobilized tosyl- (**36**, R^4 = Ts) and phenylureas (**36**, R^4 = Ph) leading to **37**. The optimized heterocyclization conditions involved the use of AuCl 5 mol% in a mixture of solvents (DCM:MeCN, 5:1) for 2 h at room temperature, similar to those reached in the first report described above.

A fully NMR spectroscopic analysis of the final heterocycles, including ^{15}N-HMBC NMR experiments, allowed to prove the high chemo- and regioselectivity of the heteroannulation, which proceeded exclusively via *N*-cyclization. The *N*-5-*exo*-dig manner cyclization was mainly observed, with the exception of some internal alkynes where the *N*-6-*endo*-dig manner was predominant, producing the zwitterion species **39** (Fig. 3). Similarly, the extension of the linker *m* (m = 2) produced the same kind of 6-membered rings.

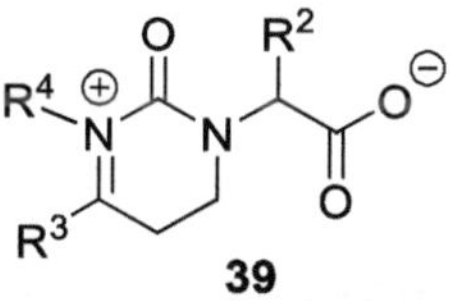

Fig. 3 6-membered zwitterionic heterocycles **39**

2.4 *Iron*

Even though the organoiron chemistry was started long time ago in 1891 by the discovery of pentacarbonyliron, iron has not played a dominant role in the catalysis development so far [84]. More recently, a remarkably increasing number of reactions using catalytic amount of iron indicate a renaissance of this metal in catalysis [85, 86]. The employment of iron catalysts is attractive for a number of reasons. Iron is the most abundant metal in earth's crust (after Al), making this metal cheaper than the precious ones. As iron compounds are present in biological systems, it has been observed that the Fe species usually exhibit relatively low toxicity. These features make this catalyst interesting for its application in several industrial areas such as pharmaceutical, food, and cosmetics. Iron can adopt oxidation states from −2 to +6 (rarely), which makes it operative as iron-centered nucleophiles apart from the most common use as Lewis acid [85]. The most frequent oxidation states for Fe are +2 and +3, being Fe(III) the most stable and widespread iron species. In addition, Fe is able to transfer both one and two electrons, enhancing the spectrum of catalysis mechanisms. All this potential explains why iron applications in organic synthesis are abruptly expanding and it is predicted that it will continue with the same tendency in the coming years. Although its expansion has begun, Fe has been yet scarcely utilized in SPOS. The reports regarding the combination of iron catalysis and solid-phase synthesis were focused on the preparation of diverse substituted [1,3,4]-thiadiazoles and dated more than a decade ago (Scheme 14) [87, 88]. In these cases, immobilized thiosemicarbazone intermediates **40** were treated with $FeCl_3$ solution to promote the cyclization towards thiadiazoles **41**. The reaction conditions for this oxidative cyclodehydratation involved simply the presence of Fe(III) in a DCM:MeOH solution at room temperature overnight, leading to the desired final heterocycles in good global yields.

2.5 *Magnesium*

Organomagnesium compounds are one of the first metal-containing reagents used in synthesis. The more common (and older) is the Grignard reagent, named after Victor Grignard, who won a Nobel Prize in Chemistry in 1912. Soon, Grignard reagent became one of the most important tools for the formation of carbon–carbon bonds. Numerous reports have been published on synthetic and industrial

Scheme 14 Synthesis of immobilized [1,3,4]-thiadiazoles **41**

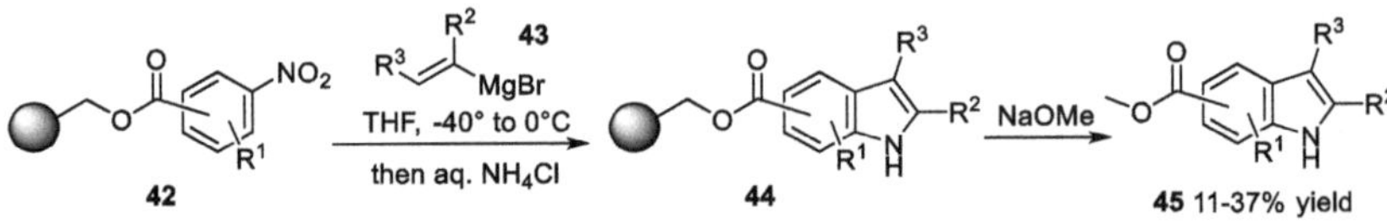

Scheme 15 Solid-phase indole synthesis under Bartoli procedure

applications of this very popular reagent [89]. However, in solid-phase synthesis of heterocycles, there are just a few examples. A couple of them describe an "in situ" heterocyclization, after treatment of an immobilized substrate with a Grignard reagent. In the first case, a Bartoli indole synthesis on solid phase was described by Bräse [90]. According to the original Bartoli procedure, excess of vinylmagnesium bromide reacts with 2-substituted nitro aryl derivatives leading to the indole moiety. One of the equivalents is used to reduce the nitro group and a second one performs the cyclization. A series of nitrobenzoic acids were attached to Merrifield resin (**42**) and treated with alkenyl Grignard reagents (**43**) to afford the indoles **44** which were detached from the polymer through a transesterification with 30% sodium methoxide (Scheme 15). Although purity was high, yields of final heterocycles **45** were generally low due to some premature release from the resin. Additionally, allylmagnesium bromide failed to give corresponding indoles.

In a later work, Bräse and coworkers took advantage of the Grignard reagents to build a library of benzobutyrolactones by a cleavage/cyclization approach [91]. Immobilization of *O*-formyl benzoic acid to Merrifield resin (**46**), followed by treatment with Grignard reagents at low temperature (−70°C), gave the benzobutyrolactones (phthalides) **48** albeit in low yields but in high purities (Scheme 16). Reaction at higher temperature mostly led to an addition reaction to the ester, producing the polisubstituted diols **49**. In the case of using ethyl magnesium chloride, reduction of the formyl group was observed. As might be expected, lithium reagents such as MeLi, EtLi, or BuLi led to premature cleavage even at −70°C. Immobilized *ortho*-formyl benzoic acids (**46**) were also used as substrate for zinc and titanium-mediated cyclizations (as mentioned below) [91].

Magnesium catalysis has been used in a very interesting approach to the synthesis of complex polyheterocyclic structures [92]. During the development of the concept of convergent DOS of small-molecule hybrids, Schreiber and coworkers reported a three-component Williams reaction [93] to obtain a sublibrary of spirocyclic oxindoles **54** that were used to build hybrid structures such as **55** (Scheme 17). Benzaldehydes tethered to a diisopropylsilyl-functionalized resin (**50**) were treated with 5,6-diphenylmorpholin-2-one (**51**) and substituted oxindolylidenes (**52**) using magnesium perchlorate as Lewis acid. The complex

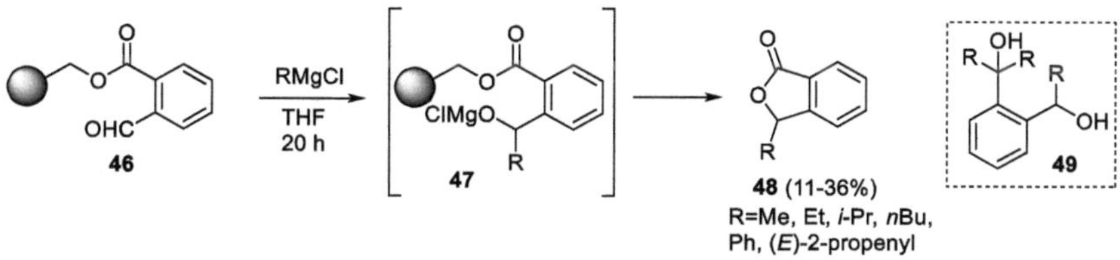

Scheme 16 Synthesis of benzobutyrolactones (phthalides) **48** using Grignard reagents

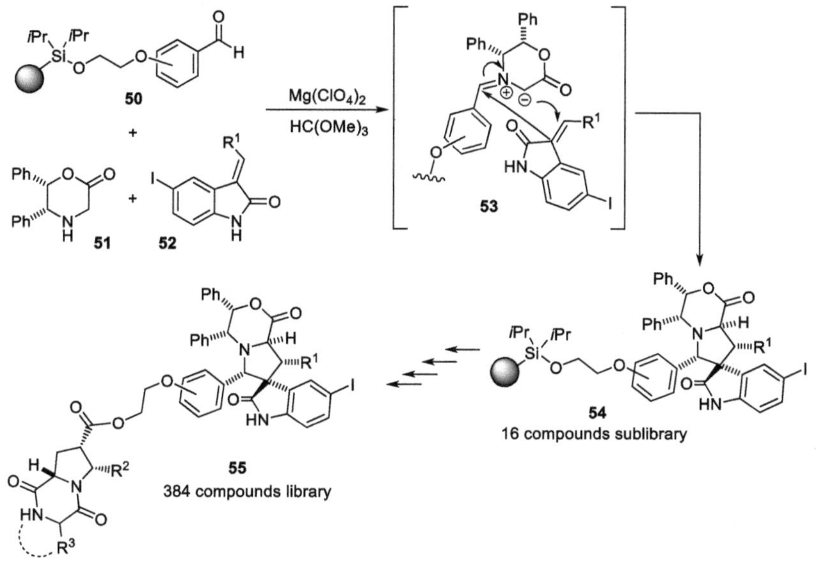

Scheme 17 Preparation of hybrid compounds library **55** from spiro compounds **54**

spirocyclic system **54** was obtained through an asymmetric [1,3]-dipolar cycloaddition of the azomethine intermediate **53**.

2.6 Mercury

Some of the most common applications of the organomercurials include the direct mercuration of aromatic compounds and the solvomercuration of alkenes [94]. The main disadvantage of organomercurials is related to their toxicity, so they must be manipulated strictly and carefully. With this precaution in mind, the organomercurials are extensively used as reagents or intermediates in organic synthesis due to many advantages such as extreme stability up to temperatures above their melting points, stable storage in air conditions, stability to protic solvents, and a substantial number of this kind of mercurated compounds commercially available [95]. Fehrentz et al. described the synthesis of 3,4,5-trisubstituted 1,2,4-triazoles

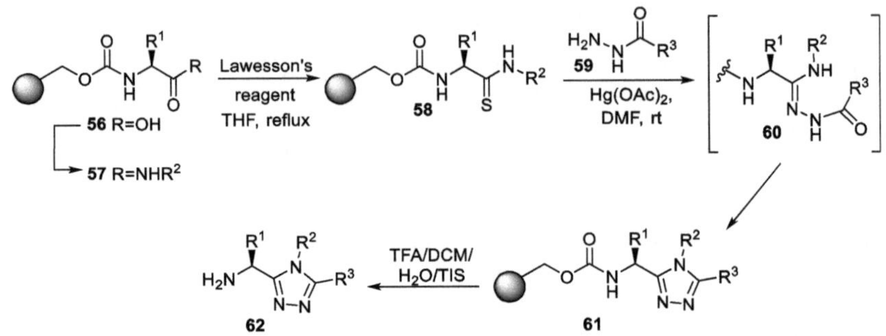

Scheme 18 Synthesis of 3,4,5-trisubstituted 1,2,4-triazoles using Mercury(II) acetate

62 as a strategy for developing libraries of peptidomimetic scaffolds [96]. Mercury (II) acetate was used as a thiophile metal salt to assist coupling-cyclization process between immobilized thioamides **58** and substituted acyl hydrazides **59** leading to heterocycles **61** under modified Takeya conditions (Scheme 18) [97]. Alfa-amino acids linked to Wang resin through a carbamate function (**56**) reacted with different amines to give **57**, and precursor thioamides **58** were performed by treatment with Lawesson's reagent. Finally, reaction with excess of both acyl hydrazides **59** and $Hg(OAc)_2$ at room temperature during 3 days afforded, after cleavage, the 3,4,5-trisubstituted 1,2,4-triazoles **62** in good purity.

2.7 *Palladium*

This metal has several features that make the reactions involving Pd catalysts and reagents particularly useful and versatile among all the transformations mediated by transition metals. Firstly, Pd catalysts offer an abundance of possibilities for carbon–carbon bond formation [98]. In addition, Pd reagents and catalysts have a remarkable tolerance to many functional groups such as carbonyl and hydroxyl groups; at the same time, the Pd species are not very sensible to oxygen neither moisture. It is also worthy to mention that the palladium toxicity is acceptable as long as working carefully [99]. Therefore, palladium is one of the most important synthetic tools in modern organic chemistry which remains under continuous development [98, 100]. The most commonly used palladium compounds in organic synthesis are Pd(II) salts and Pd(0) complexes. Pd(0) complexes are employed as catalysts while Pd(II) compounds are mainly used as oxidizing reagents (only few examples reported) and as precursors of Pd(0) species.

Similarly to solution chemistry, palladium has been extensively employed in SPOS in a plethora of chemical transformations [101]. Various Pd-mediated cyclizations on solid support have been efficiently developed towards different heterocyclic skeletons, as it will be described hereafter. Certainly, the most common heterocycles provided by palladium catalysis in SPOS are the indolic scaffolds, thus

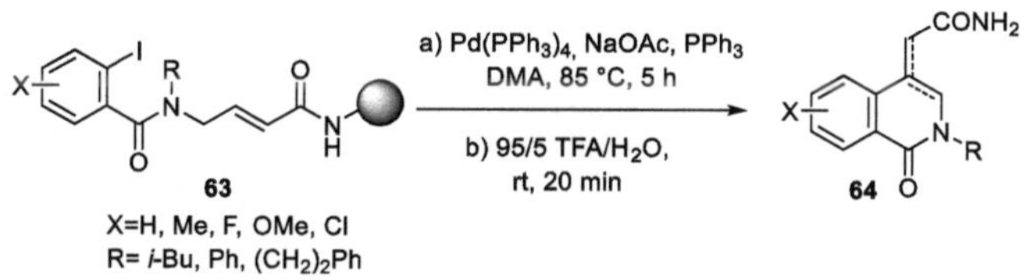

Scheme 19 Synthesis of 2*H*-isoquinolinones **64** catalyzed by Pd(0)

they will be detailed separately in the following Sect. 2.7.1. The first report referred to a supported cyclization mediated by palladium dated in 1995 and described the synthesis of 2*H*-isoquinolinones **64** from resin-bound monopeptoids **63** (Rink amide resin) (Scheme 19) [102]. Afterwards, a few reports have described the synthesis of isoquinolines via Pd catalysis on solid support (using Rink resin [103] and an additional base-labile linker polystyrene resin [104]). Some of the isoquinolines underwent slow air oxidation to the corresponding isoquinolines [104].

Since the mid-1990s, the use of palladium for the generation of immobilized heterocycles expanded notably towards very versatile scaffolds. Similarly to previous cyclization, many strategies started from immobilized aryl halides. In some cases, the combination of these substrates with alkenyl and alkynyl chains allowed the cyclization via Heck type reaction. This methodology was used for the construction of 7-membered rings, included in heterocyclic molecules with interesting potential biological activities. For instance, the synthesis of substituted benzazepines **66** was carried out via intramolecular 7-*exo* Heck cyclization from supported iodo benzyl precursors **65** on Wang resin (Scheme 20a) [105]. An extension of the methodology to produce this kind of interesting bicyclic lactams was carried out via a Ugi/Heck sequence and MW-assisted Pd cyclization [106]. In a similar fashion, aryl halides coupled to cysteine (**67**) allowed 7-membered ring closures to provide benzofused sultams **68** via palladium-catalyzed Buchwald–Hartwig type intramolecular cyclization (*p*-methylbenzhydrylamine resin) (Scheme 20b) [107, 108]. These libraries were diversified by further modifications at the nitrogen present in the benzofused cyclic sulfonamides in **68**, as well as by the incorporation of an amide functionality instead of the sulfonamide group, leading to the corresponding benzothiazepinone scaffolds [108].

Benzofurans and their derivatives and analogues have been prepared from immobilized iodo phenol substrates via Sonogashira reaction followed by Pd-catalyzed annulation by several research groups with different purposes. The first report describing this methodology was published by Bedeschi et al. in 1997 [109]. The TentaGel resin-bound *ortho*-hydroxy aryl iodide **69** smoothly reacted in the Pd-catalyzed heteroannulation step, subsequent to the alkyne incorporation, to provide the benzofuran nucleus **70** (Scheme 21). The use of the base tetramethylguanidine (TMG) has been reported to allow good yields and milder reaction conditions. Similar strategic concept was applied to the preparation of a 210-membered 2-substituted 3-arylbenzo[*b*]furan library, using $Pd_2(dba)_3$

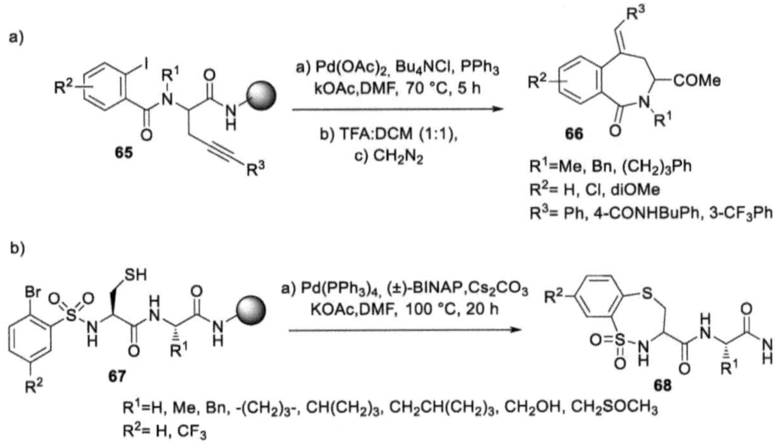

Scheme 20 Synthesis of 7-membered heterobicycles: (**a**) benzazepines **66** via Heck cyclization and (**b**) benzofused sultams **68** via Buchwald–Hartwig type cyclization

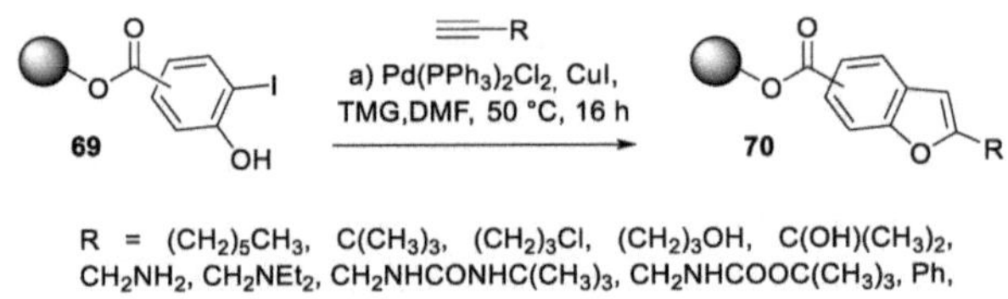

Scheme 21 Synthesis of 2-substituted benzofurans **70** via Sonogashira reaction followed by palladium-catalyzed heteroannulation of acetylenes

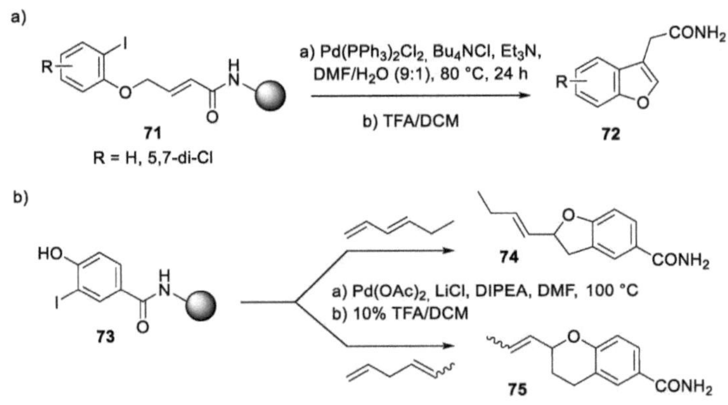

Scheme 22 Benzofurans **72**, 2,3-dihydrofurans **74**, and 2,3-dihydrobenzopyrans **75** prepared by Pd-catalyzed annulation

(2 equiv.) and bipyridine (4 equiv.) [110]. In analogy, resin-bound 3-propargylamino-2-seleno-ester (organoselenium resin) reacted with 2-iodophenol under Pd/Cu conditions to provide a bis-heterocycles library of

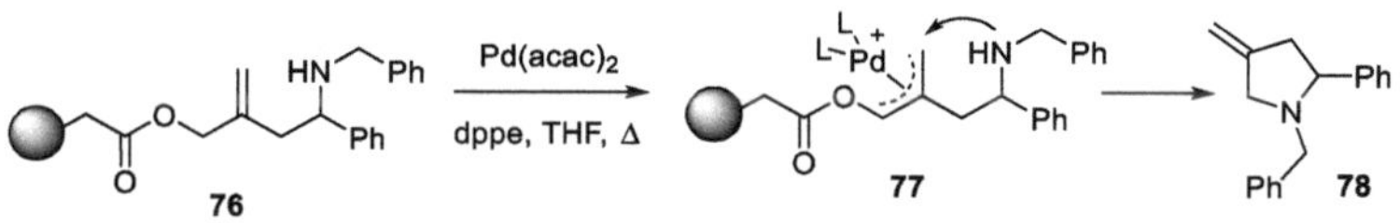

Scheme 23 Cyclization and cleavage in the solid-phase synthesis of pyrrolidines described by Brown

uracil/diazepinedione and benzofurans, via a Sonogashira/heterocyclization sequence [111]. The use of 2-iodoanilines reagents has been amply explored for the generation of indolyl nucleus as it will be described in the following section.

In addition, alternative sequences for preparations of benzofuran derivatives were carried out coupling aryl halides with alkenes. Maryanoff and coworkers prepared benzofurans **72** from Rink amide resin-bound *O*-alkenyl iodofenol substrates **71** via a Pd-promoted intramolecular Heck type reaction (Scheme 22a) [112]. Briefly after, similar strategy was employed for the synthesis of heterocyclic hydroxamates, including dihydrobenzofurans, via an acylpalladium(II)-catalyzed termolecular cascade reaction involving alkenyl-substituted iodo benzene derivatives, carbon monoxide, and protected hydroxylamines [113]. In addition, the cyclo-coupling of dienes to immobilized 2-iodophenols **73** mediated by $Pd(OAc)_2$ led to 2,3-dihydrofurans **74** and 2,3-dihydrobenzopyrans **75** (Scheme 22b) [103].

Considering the solid-phase synthesis of monoheterocycle structures that include palladium-catalyzed annulation, the synthesis of pyrrolidines, published by Brown, is an example [114]. In this paper, an interesting linker is described. Under palladium catalysis, electrophilic species are generated which could be trapped by a heteroatom to release the heterocyclic compounds from the resin (Scheme 23). Based on this, a small solid-phase library of pyrrolidines was synthesized from the key immobilized homoallylic amine **76**, that underwent palladium catalytic cyclization and cleavage, to afford pyrrolidines as **78** in moderate yields.

Butenolides are present in a large number of natural products. Albeit there are many published solution syntheses of these compounds, and they are difficult to automate. To fix this, in 2000 Ma et al. published a solid-phase synthesis of butenolides [115]. To carry out this strategy, *p*-/*o*-iodobenzoic acid, *p*-iodophenol, or *p*-iodo benzyl alcohol were anchored at Merrifield resin to afford different supported iodoaryl compounds (**79a–d**, Scheme 24), that reacted with 1,2-allenyl carboxylic acids **80** under palladium catalysis generating the heterocycles **81**. Several cleavage conditions were tested, being a mixture of acetyl bromide and zinc(II) bromide the best option found. Butenolides **82** were obtained in high yields and purities from the crude mixture.

Palladium has also promoted macrocyclization processes, leading to cyclic tetrapeptides via Heck reaction in a more efficient manner than the corresponding transformation in solution [116].

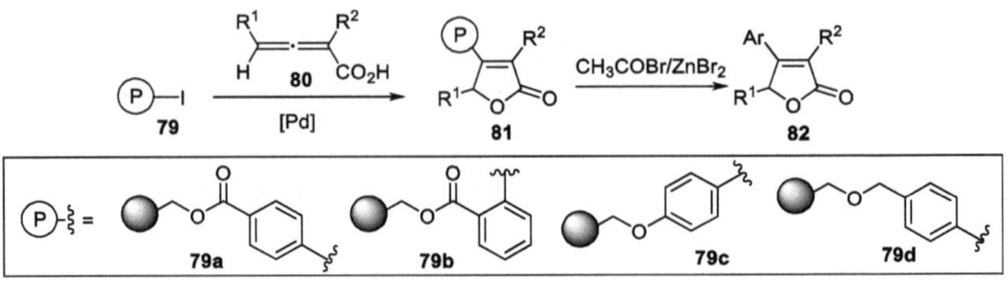

Scheme 24 Solid-phase synthesis of a butenolides library

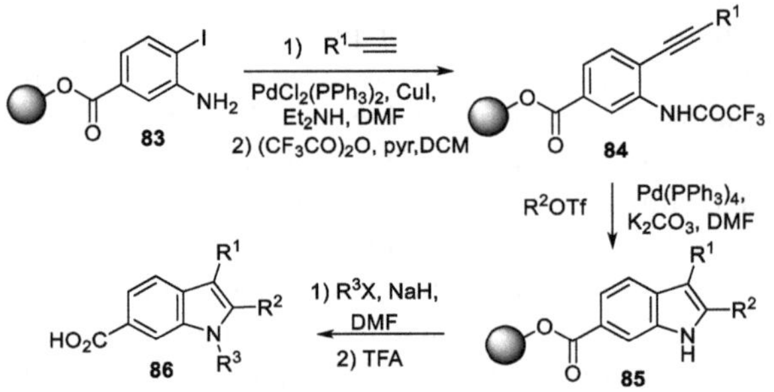

Scheme 25 Main stages in the solid-phase synthesis of trisubstituted indoles

2.7.1 Synthesis of Indolyl Nucleus Mediated by Palladium

Indolic rings are units very common in compounds with biological activity, and so the availability of solid-phase strategies for the synthesis of libraries of these derivatives is highly desirable. Many of the approaches involve the use of Pd-catalyzed reactions for the heterocyclic generation. Based on this, a combinatorial library of indoles was designed in 1997 [117], starting from Wang-supported 3-amino-4-iodobenzoic ester **83** (Scheme 25). In the generation of **85**, two sequential Pd-catalyzed transformations were involved (introduction of the R^1 and R^2 groups). For that, a cross-coupling under Sonogashira conditions occurs first, and then the incorporation of a vinyl group (R^2 = –CH=CHR) and cyclization using K_2CO_3 as base. Finally, different multisubstituted indoles **86** were obtained after *N*-alkylation and acid treatment (33–81% isolated yields).

Other analogous solid-phase methodologies involving palladium catalysis but in which cross-coupling and cyclization occur in a single step have been also reported. Bedeschi et al. [118] linked, under Mitsunobu conditions, 3-iodo-4-acetamidobenzoic acid to TentaGel-STM resin to provide **87** (Scheme 26). Then **87** was coupled with the alkyne **88**, and the resulting compound cyclized "in situ" to indole **89**. The efficiency of this one-pot transformation was attributed to the ability of TMG base to promote the coupling and the cyclization. Indoles **90** were obtained from **89** after basic (NaOH) or nucleophilic conditions (NaOMe/MeOH). Yields were determined by quantitative HPLC analysis (48–95%).

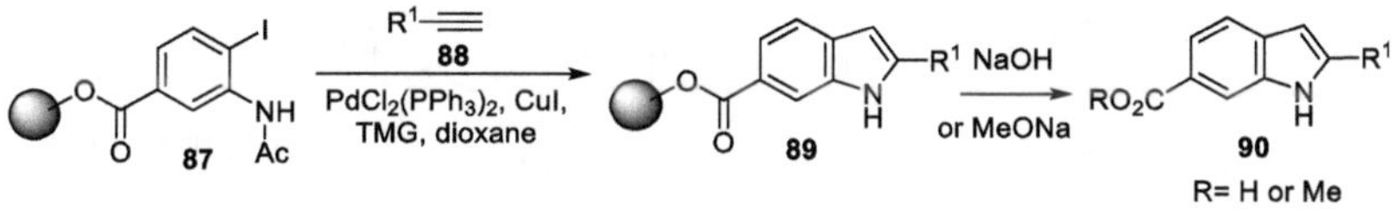

Scheme 26 Solid-phase indole synthesis by one-pot coupling and cyclization

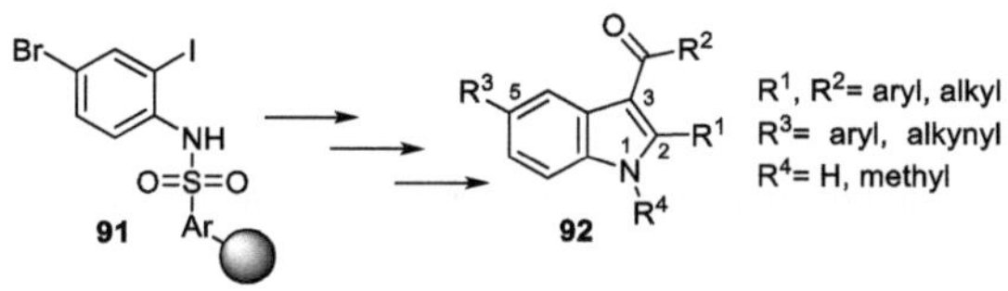

Scheme 27 Solid-phase synthesis of a 2,3,5-trisubstituted indole library

2,3,5-Trisubstituted indoles were synthesized starting from polystyrene sulfonyl resin-bound aniline **91** (Scheme 27) [119]. In this strategy, R^1 group was incorporated at **91** through a palladium-catalyzed coupling with several terminal alkynes and then an intramolecular cyclization afforded the indole core. Then, acylation by acid catalysis introduced R^2 at C-3 position. Finally, the substituent on the C-5 was modified by Sonogashira or Suzuki cross-couplings. Cleavage of the resin-bound 2,3,5-trisubstituted indoles was performed by the treatment of *t*-BuOK, giving the soluble indoles **92** in overall yields from 10 to 20%.

Several structurally complex indoline libraries were constructed using solid-phase mix and split methodologies [120]. For this purpose, palladium-catalyzed tandem reactions were used (cascade Stille cross-coupling-cyclization-anion capture and cascade Suzuki cross-coupling-cyclization-anion capture). Wang resin-bound aryl iodide reacted with vinylstannanes (Stille conditions) or alkenylborates (Suzuki conditions) furnishing indole structures. Based in this strategy, three indole libraries have been prepared in excellent yields.

A chiral version synthesis of indolines was developed by Houghten [121], starting from resin-bound acylated amino acids as chiral components and using the "tea-bag" solid-phase methodology. In this case, **93** was reduced to **94** which was, in turn, converted to the indole core under palladium intramolecular catalysis (Scheme 28). Finally, **95** was obtained in good yields, purity, and without racemization, after cleavage from the resin using HF/anisole (95/5).

Huang et al. published in 2010 [111] the solid-phase synthesis of libraries of bis-heterocycles looking for compounds with potential and synergistic bioproperties. The bis-heterocycle systems were composed of uracil/diazepinedione and benzofuran/indole units (Scheme 29a). First, resin-bound 3-amino-2-seleno-ester **96** was transformed into uracil or diazepinedione by treatment with isocyanate, or R-aminoacid followed by basic cyclization (Scheme 29b). Then supported alkynes **99** were transformed to benzofuran or indole through a Sonogashira/annulation reaction with 2-iodophenol or 2-iodoaniline (Scheme 30). Finally the compound **102** was released from the resin prior to oxidation and

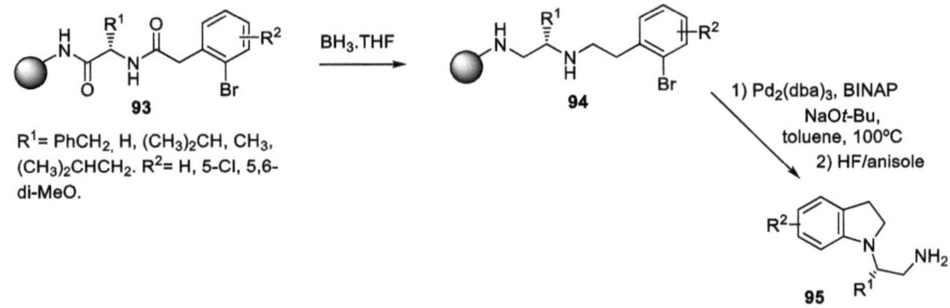

Scheme 28 Chiral version of solid-phase syntheses of indoline libraries

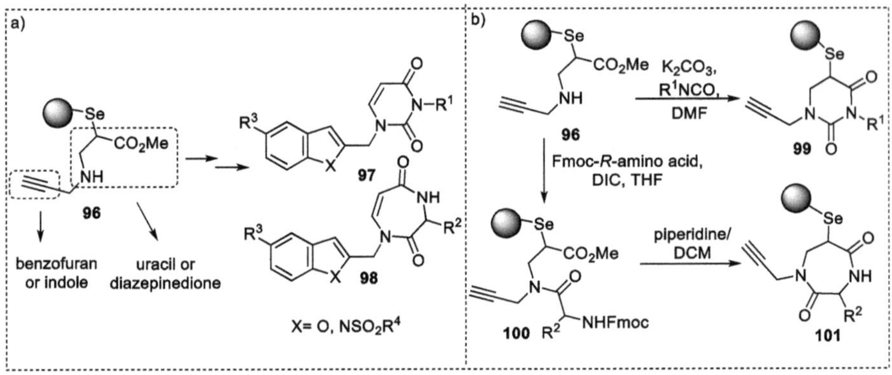

Scheme 29 (**a**) General outline of bis-heterocycle synthesis and (**b**) synthesis of **99** and **101**

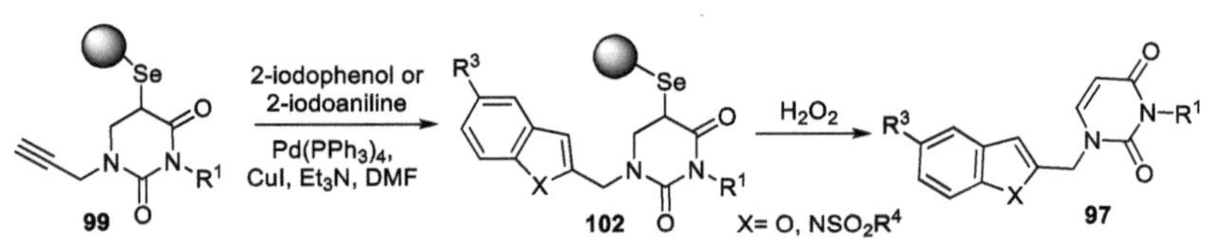

Scheme 30 Solid-phase synthesis of bis-heterocyclic systems composed by benzofuran or indole and the uracil units

elimination of selenium linker. The same sequence was applied on **101** to afford the system composed by the benzofuran or indole and the diazepinedione unit **98**.

Kondo et al. published a series of papers of solid-phase synthesis [122–124] involving cyclization reactions catalyzed by palladium for the generation of libraries of indole carboxylates. For the synthesis of immobilized substrates **105** and **107** (Scheme 31), three different alternatives were developed: (1) Rhodium-catalyzed N–H insertion via a supported-diazo compound **103** to afford **104**, which was transformed into **105** by treatment with an aldehyde under Horner–Emmons conditions (Method A). (2) Treatment of the α-ketoester **106** with 2-haloanilines (Method B), or (3) by oxidative amination of **108** with 2-haloanilines under

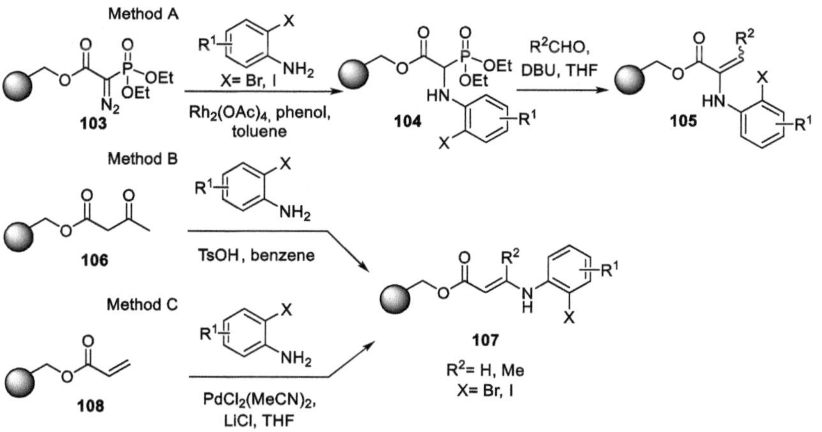

Scheme 31 Synthetic alternatives for the generation of starting materials **105** and **107** for the preparation of indoles described by Kondo

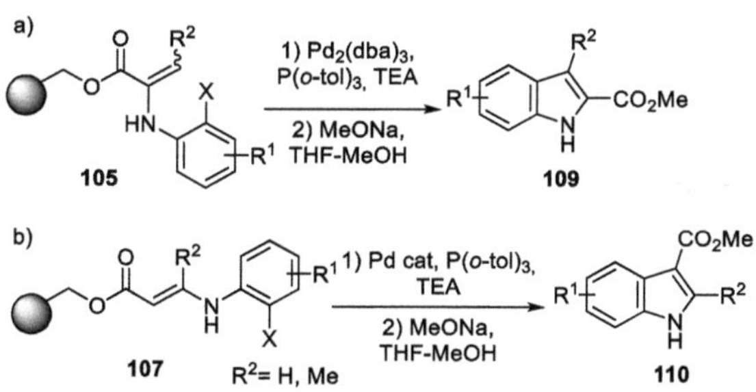

Scheme 32 Palladium-catalyzed cyclization of indole carboxylates **109** and **110**

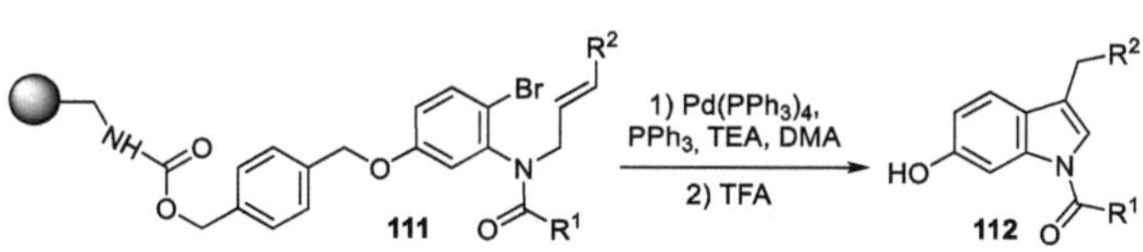

Scheme 33 Synthesis of indol analogues from immobilized allylamide **111**

palladium catalysis (Method C) both to give **107**. Finally, with these starting materials, palladium-catalyzed intramolecular cyclization was carried out under the conditions shown in the Scheme 32 giving, after transesterification (MeONa in MeOH–THF), the methyl carboxylates **109** and**110** (31–78% isolated yield). Although for the synthesis of **109**, $Pd_2(dba)_3$ was the best option (Scheme 32a), in the case of the synthesis of **110**, $Pd_2(dba)_3$ or $Pd(OAc)_2$ were used optionally, depending on the substrate substituents (Scheme 32b).

Indol analogues were also synthesized by intramolecular cross-coupling Heck reaction, starting from TentaGel-supported allylamide **111** (Scheme 33) [125]. This

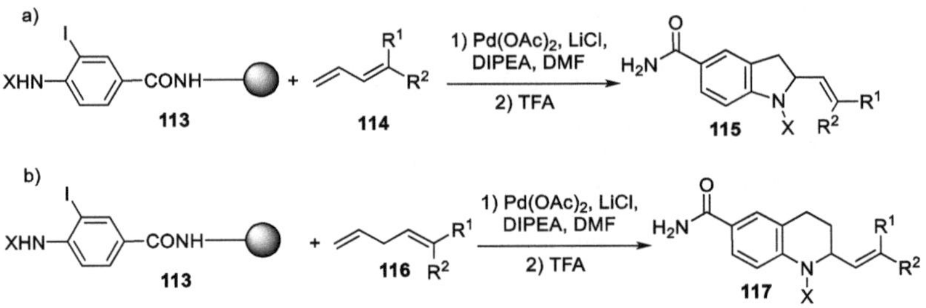

Scheme 34 Solid-phase synthesis of indolines (**115**) and 1,2,3,4-tetrahydroquinolines (**117**) by palladium-catalyzed annulation

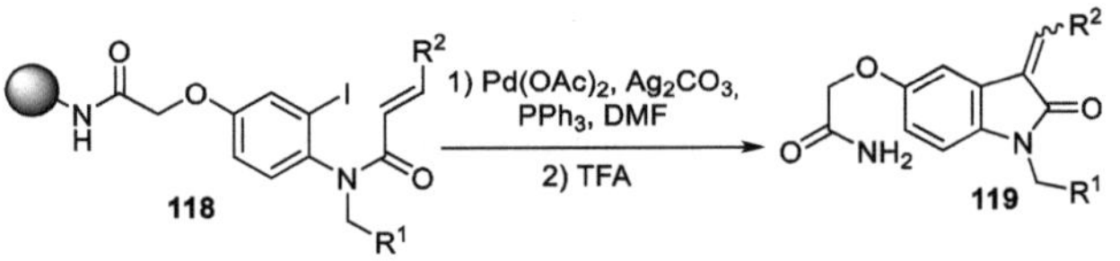

Scheme 35 Solid-phase synthesis of 2-oxoindole library

compound, by treatment with a catalytic amount of $Pd(PPh_3)_4$, PPh_3, and anhydrous *N*,*N*-dimethylacetamide (DMA), gave **112** in a 65–94% crude yield, after TFA cleavage. A one-pot cross-coupling and cyclization strategy is shown in Scheme 34a [103]. Using Rink resin, the cyclization was performed by heating **114** with different 1,3-dienes, $Pd(OAc)_2$, LiCl, and DIPEA in DMF at 100°C. Acid cleavage with 10% TFA allowed the release from the resin, providing **115** in good yields and high purities. Alternatively, when 1,4-dienes **116** were used, 1,2,3,4-tetrahydroquinoline derivatives **117** were obtained (Scheme 34b). Using this strategy, high diversity could be generated; for example, when solid-supported *o*-iodophenol was employed instead of iodoaniline in the palladium-catalyzed annulation, 2,3-dihydrobenzofuran and 2,3-dihydrobenzopyran derivatives were synthesized, as previously described.

Intramolecular Heck cross-coupling has also been applied for the construction of 2-oxindole derivatives [126]. By applying Heck conditions on Rink amide resin-supported iodide **118**, different 3-ethylidene-2-oxindoles **119** were generated (Scheme 35). The 2-oxindole library was separated from the resin by acid treatment and analyzed by ^1H NMR and HPLC, showing the presence of (*E*)- and (*Z*)-isomers.

Zhang et al. have published a series of works in which the solid-phase synthesis of highly substituted indoles were reported [112, 127–130]. Supported iodoanilines **120** (Scheme 36) were treated with excess of several commercial alkynes, palladium acetate (10 mol%), Ph_3P (20 mol%), LiCl, and K_2CO_3 in DMF at 80°C, giving good yields of the indoles **121**. When unsymmetrical alkynes were evaluated, the major product was the 2,3-disubstituted indoles having the larger substituent at the 2-position. Regioselectivity was attributed to the insertion of the palladium species,

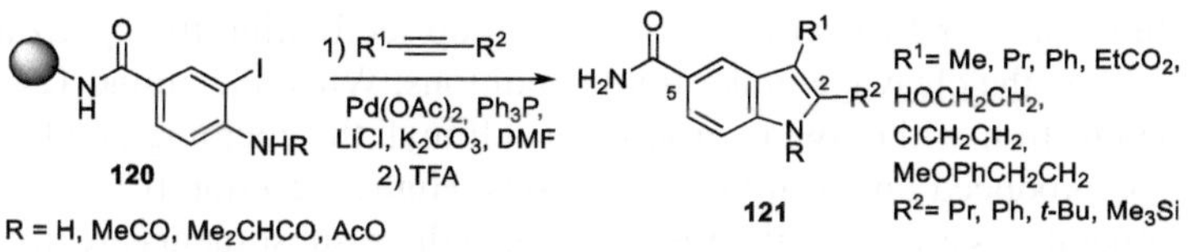

Scheme 36 Solid-phase synthesis of indoles catalyzed by palladium acetate, immobilized by position 5

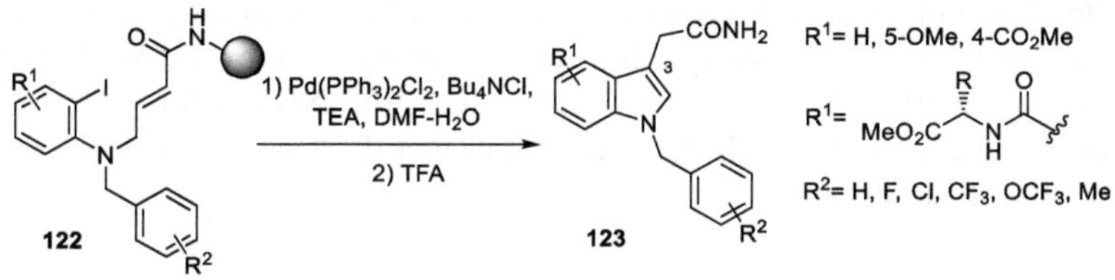

Scheme 37 Synthesis of indoles by palladium catalysis, immobilized by position 3

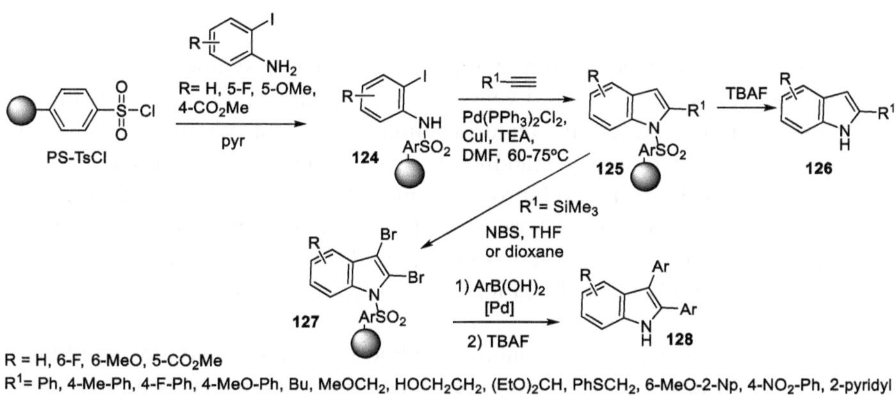

Scheme 38 Synthesis of indoles by traceless linkers

during the oxidative addition, to the less hindered side of the alkyne, placing the small substituent at the 3-position. The reaction was performed with a wide variety of alkynes (shown in Scheme 36). Crude yields were 38–100% and purities 53–94%.

Immobilization of the indole by its C-3 position was also evaluated (Scheme 37). Substrate **122,** attached to Rink amide resin, underwent a palladium intramolecular cyclization under phase-transfer conditions, releasing indole **123** after treatment with TFA. Very good yields were obtained after purification. Through this methodology, a chiral version was developed by using substrates in which a chiral amino acid was incorporated in R^1. This strategy was also extended to the development of libraries of benzofurans, as previously mentioned [112].

Another possibility studied by this research group was the solid-phase synthesis of indoles using traceless sulfonyl linkers (Scheme 38). They suggest that the linker has two functions: first to act as activating group facilitating the cyclization, and second to allow the cleavage under mild conditions. When the amine **124** is linked to a strong electron-withdrawing group as a sulfonyl, the coupling and the cyclization may be produced under milder conditions. Higher reaction times and slightly higher temperatures are generally required when these substituents are not used. To the synthesis, PS-TsCl or Merrifield resin-based sulfonyl chloride were employed. Cross-coupling Sonogashira conditions were applied, to immobilized amine **124** at varying temperatures and reaction times depending on the substituent of the starting substrate. The indole library (**126**) was generated in very good yields and purity (85–100% yield, 85–100% purity) after removal of the resin by TBAF. When R^1 in **125** was a trimethylsilyl, treatment with excess of NBS afforded the 2,3-dibromoindole **127**, from which the Suzuki product **128** was obtained, expanding the diversity of the library.

2.8 *Rhodium*

The application of organorhodium compounds in organic synthesis has shown an increasing development in the last years. Such compounds catalyze a large number of transformations, generally featuring high levels of chemo-, regio-, and stereoselectivity, such as cycloadditions, C–H insertion, generation of metal carbenoids useful to afford cyclopropanes and other derivatives, etc. Being these reactions interesting for the generation of heterocyclic compounds, an extensive development of this chemistry has been reported, mainly in homogeneous phase [131], while the number of publications using supported substrates has been much lower.

In 1997, two research groups published almost simultaneously, the solid-phase synthesis of furans via 1,3-dipolar cycloaddition through rhodium catalysis [132, 133]. Gallop et al. [132] described the reaction of TentaGel amine resin-bound α-diazocarbonyls **130(a–d)** (Scheme 39), generated from **129(a–d)**, with electron-deficient acetylenes under Rh(II) catalysis to afford under thermal conditions, furan derivatives in a traceless linkage strategy. At the end of the sequence, only the products **136** to **144** and an excess of unreacted acetylene were observed. The products were purified by column chromatography being obtained in 50–70%

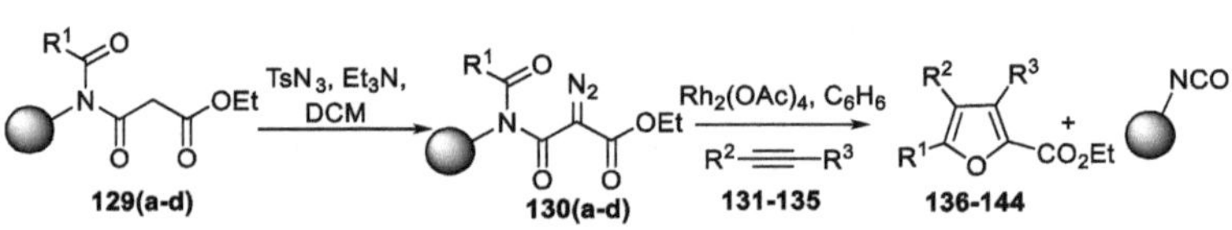

a) R^1=CH_3; b) R^1=CH_3CH_2Ph; c) R^1=4-CH_3O-C_6H_4, d) R^1=$^{13}CH_3$

Acetylenes: (**131**) dimethyl acetylenedicarboxylate (DMAD); (**132**) dibenzoylacetilene; (**133**) diethyl acetyl-enedicarboxylate (DEAD); (**134**) methyl propiolate; (**135**) ethyl propiolate

Scheme 39 Solid-phase synthesis of furans by rhodium catalysis: Gallop methodology

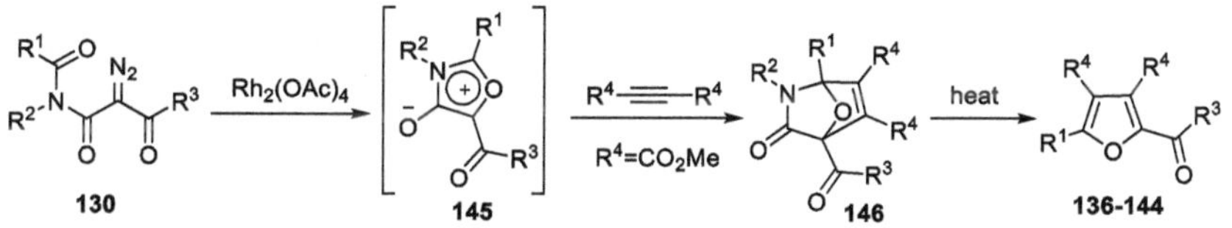

Scheme 40 Mechanistic proposal to the generation of furans via rhodium catalysis

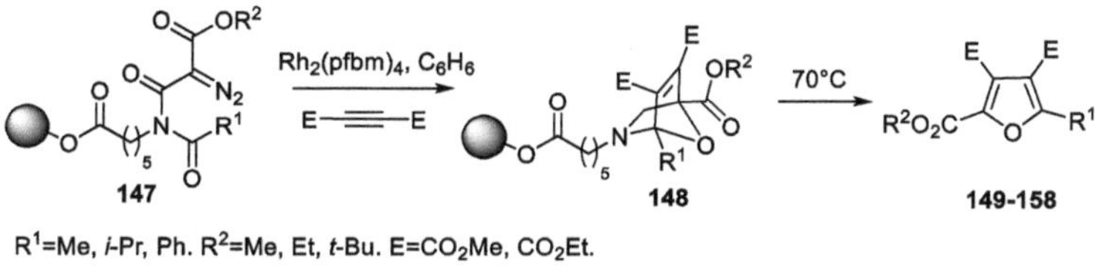

Scheme 41 Solid-phase synthesis of furans by rhodium catalysis: Austin methodology

yields. Furthermore, using this methodology, one small combinatorial library of furans was generated (32 member) via split synthesis.

According to the proposed mechanism (Scheme 40), the α-diazocarbonyls suffer an intramolecular reaction giving isomünchnones (**145**), via a rhodium carbenoid. The presence of acetylenes leads to a [2 + 3] cycloaddition affording bicyclic intermediates (**146**) which undergo a cycloreversion losing isocyanate and generating the corresponding furans (**136–144**).

Austin et al. published a series of works quite similar to the previous one [133, 134]. Very good yields and purities were obtained, following the synthetic sequence shown in Scheme 41. In this strategy, the Wang resin-supported diazo derivative **147** underwent a cycloaddition with dialkylacetylene dicarboxylates at room temperature, using rhodium(II) perfluorobutyramide ($Rh_2(pfbm)_4$) as catalyst, to afford immobilized bicycles **148**. The compounds **148** can be thermally cleaved from the solid support (benzene, 70°C) to afford the corresponding furans **149–158**. The advantage of this strategy comparing to the previous one (Gallop's methodology) is that the unreacted alkyne and the catalyst can be removed by filtration prior to the cycloreversion cleavage step, giving the furans **149–158** in higher purity.

Related with this work, in 2002 the same group published an asymmetric solid-phase synthesis of [2.2.1] bicyclic scaffolds [135]. In this article, D- and L-hydroxyvaline were linked to benzhydrylamine (BHA) resin to afford the immobilized auxiliaries (*S*)-**160** and (*R*)-**160** which, in turn, were treated with *N*-acetyl-*N*-methylmalonamic acid and subjected to diazotransfer to give the chiral-supported diazo derivatives (*S*)-**161** and (*R*)-**161** (Scheme 42). Treatment of (*S*)-**161** or (*R*)-**161** in the presence of different vinyl ethers and $Rh_2(pfbm)_4$ as catalyst gave bicyclic structures **162–168**, after resin removal employing methylamine.

Following the applications of rhodium carbene additions to solid-phase methodologies, supported synthesis of oxazoles derivatives was carried out by the Iso's group [136]. In this strategy, the supported acid chloride **169**, generated after several steps from Wang resin, was transformed into **170** by reacting with TMS diazomethane (Scheme 43). Intermediate **170** evolved to supported oxazoles (**172**),

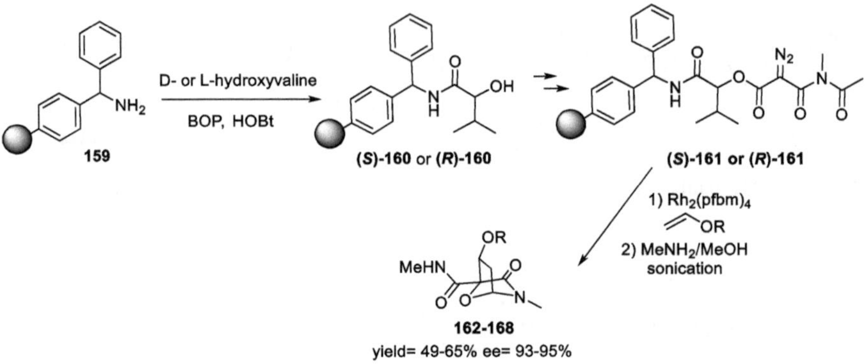

Scheme 42 Diastereofacial solid-phase synthesis of biheterocyclic compounds

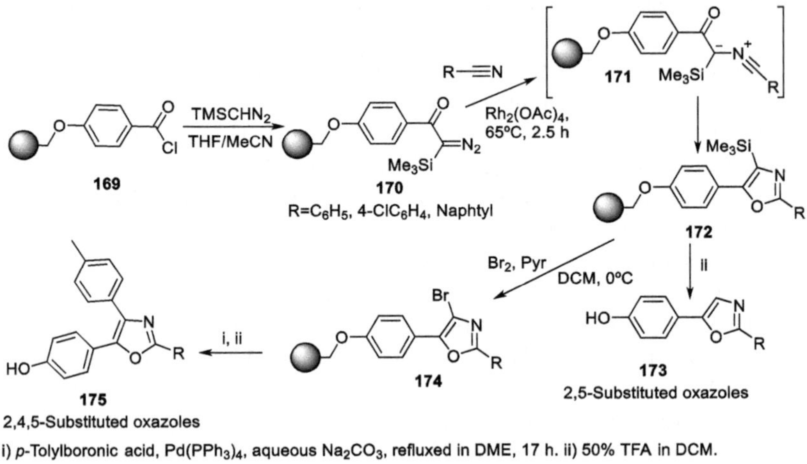

Scheme 43 Libraries synthesis of substituted oxazoles

by treatment with aryl nitriles and rhodium catalysis. The reaction proceeds via ylide formation and subsequent 1,5-cyclization. The reaction was efficient both in the presence of benzene or under solvent-free conditions. Di- (**173**) and trisubstituted oxazoles (**175**) can be generated by acid treatment of **172**, or bromination followed by a Suzuki reaction and then cleavage, respectively.

The metal-catalyzed [2 + 2 + 2] cyclotrimerization of diynes and alkynes are very useful approaches to the synthesis of highly substituted aromatic rings. As it was already mentioned, this reaction is especially suitable for solid-phase chemistry, avoiding undesired homocouplings. In 2001, Sun published a work regarding the solid-phase synthesis of isoindolines via a rhodium-catalyzed [2 + 2 + 2] cycloaddition [137]. In the key step, resin **178** was reacted with different substituted alkynes under Wilkinson's catalysis at 80°C in $CHCl_3$/EtOH to give the supported isoindolines **179** (Scheme 44). Products **180** were obtained, after cleavage from the resin by TFA treatment, in 20–85% isolated yields.

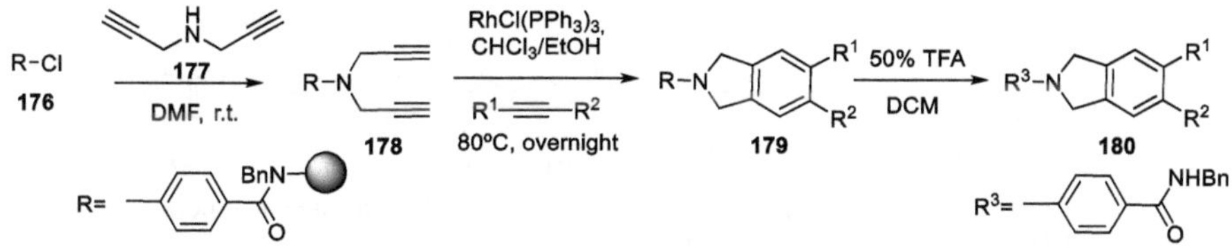

R[1]= H, Me. R[2]=CH_2OH, CH_2OMe, CH_2NMeBn, $(CH_2)_5Me$, 1-cyclohexen, Ph.

Scheme 44 Solid-phase rhodium-catalyzed [2 + 2 + 2] methodology developed by Sun

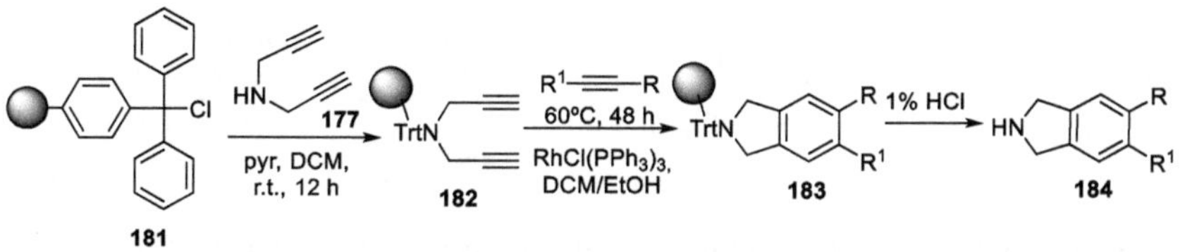

R= H, CH_2CH_3, CH_2COCH_3. R[1]=$(CH_2)_3CH_3$, Ph, CH_2OH, CH_2OBn, CH_2NBoc, $(CH_2)_3CN$, $SiMe_3$, $(CH_2)_4Cl$, CO_2CH_3, CH_2COCH_3.

Scheme 45 Solid-phase synthesis of isoindolines by rhodium-catalyzed [2 + 2 + 2] methodology

As part of their study regarding the solid-phase [2 + 2 + 2] cyclotrimerization using different metal catalysts (Co, Rh, and Ru), Dieters et al. have reported the rhodium-catalyzed generation of supported isoindolines and phtalans [138]. Trityl resin-immobilized diyne **182** reacted with a set of alkynes with different functionalities, also in the presence of Wilkinson's catalyst (Scheme 45). Acetylene was most reactive and the reaction was conducted at room temperature. Higher temperatures were needed when monosubstituted alkynes were used. Hydroxy and amine groups gave better performance when they are protected. Disubstituted alkynes led to the formation of the desired product but the yields were lower. In general, better results were found by degassing the solvents prior to catalyst addition and pouring it in two portions. Isoindolines **183** were separated from the resin by treatment with 1% anhydrous hydrochloric acid in CH_2Cl_2. The products **184** were obtained in 70–95% yields and 90% purity without further purification.

When the previous strategy was used for the synthesis of phtalans, only trace amounts of cyclotrimerization products were obtained. Switching the immobilized diyne to a compound attached to a carboxy-linked resin by one of the sides of the moiety, such as **185**, phtalans **187** were obtained in 57–82% yields (Scheme 46). The authors attribute the negative results using trityl resin to steric problems. Low or no regioselectivity was observed in these cases, the problem that was circumvented by employing Ru catalysts (see Sect. 2.9, Scheme 54).

2.9 *Ruthenium*

Ruthenium is today one of the most popular transition metals used in organic synthesis [139]. However, until the eighties, the reported useful methodologies

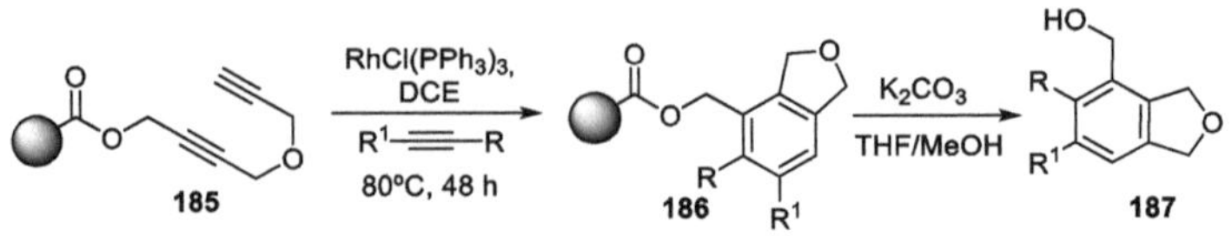

R= H, CH_2CH_3, CH_2COCH_3. R^1=$(CH_2)_3CH_3$, Ph, CH_2OH, CH_2OBn, CH_2NBoc, $(CH_2)_3CN$, $SiMe_3$, $(CH_2)_4Cl$, CO_2CH_3, CH_2COCH_3.

Scheme 46 Solid-phase synthesis of phtalans by rhodium-catalyzed [2+2+2] methodology

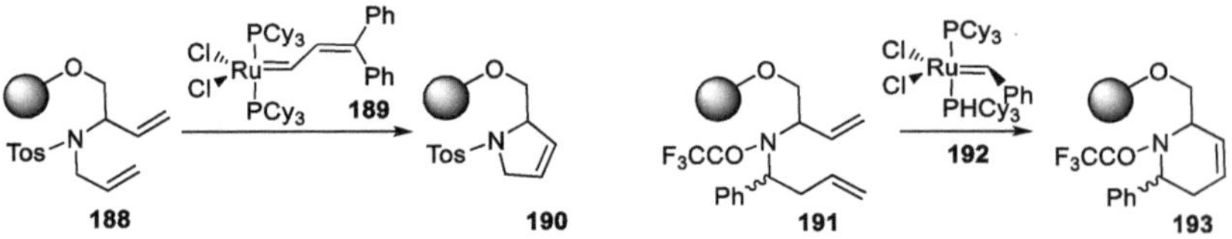

Scheme 47 Solid-phase synthesis of five- and six-membered heterocycles by RCM

involving ruthenium were comprised of a small number of reactions such as RuO_4 oxidations, hydrogenation, and hydrogen transfer reactions. The emergence of well-defined ruthenium catalysts during the last two decades, and most importantly, their application to metathesis reactions, gave this transition metal its current importance [140]. Metathesis reactions have become a powerful synthetic tool in modern organic and polymer synthesis, representing an excellent alternative to traditional carbon–carbon bond forming reactions, such as Stille, Heck, and Suzuki couplings [141]. Olefin metathesis is a reversible process consisting in an exchange of alkylidene groups between two olefins, catalyzed by metal carbene complexes, while enyne metathesis is a reorganization of an olefin and an alkyne to generate a conjugated diene [142]. For olefin metathesis, three major categories can be described: ring-closing metathesis (RCM), ring-opening metathesis (ROM), and cross-metathesis (CM). In the case of the enyne metathesis, it can be divided between the intermolecular or cross-enyne metathesis (CEM), and the intramolecular or ring-closing enyne metathesis (RCEM or RCEYM) [143].

Several authors have taken advantage of the olefin and enyne metathesis for the ring closure to generate heterocycles [144]. The first application of RCM to solid-supported chemistry was owed to Blechert et al. [145] who succinctly reported the preparation of five- and six-membered ring heterocycles either using precatalyst **189** [146] or first-generation Grubbs catalyst (**192**) [147] (Scheme 47) [148].

In addition to the well-known benefits of solid-phase synthesis, performing the cleavage simultaneously to the cyclization would save one step in the synthetic process. Most of the RCM on solid phase were achieved by this strategy. Precisely, the synthesis of a caprolactam was described by RCM using cleavage/cyclization approach [149]. Thus, Merrifield resin-immobilized diene **194** was treated with as much as 100 mol% of **192** to yield the ε-caprolactam **195** with concomitant release from the resin, in moderate yield (Scheme 48a). An improved process was later reported by the same group for the synthesis of a library of nitrogenous heterocycles

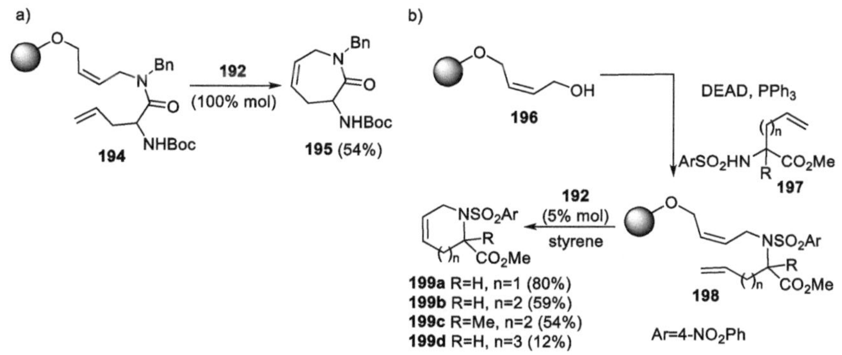

Scheme 48 Cleavage/cyclization by RCM for generating nitrogenous heterocycles

[150]. In this case, Wang resin-supported allylic alcohol **196** was treated, under Mitsunobu conditions, with different α-alkenyl glycine methyl ester derivatives **197** to obtain precursors **198** which, in turn, were subjected to the RCM with precatalyst **192**, to afford pipecolinic acid derivative **199a** in high yield for the cyclization/cleavage step (Scheme 48b). Seven- and eight-membered azepine (**199b–c**) and azocine (**199d**) derivatives were also obtained, albeit in lower yields. Interestingly, the authors found that addition of styrene during the RCM significantly improved yields, probably by helping the Ru species to re-enter in the catalytic cycle. In the presence of styrene, formation of stilbene is a logical impurity and an important drawback for the reaction.

Using a closely related approach, Piscopio and coworkers [151] obtained a pipecolinic acid derivative similar to **199a** in 62% yield without addition of any additive. Application of the procedure to the synthesis of Freidinger lactams was also described [152]. These lactam-bridged dipeptides have been widely used to constrain peptide conformations [153]. Immobilized 2,4-dinitrobenzenesulfonamide **200** was synthesized and the nitrogen was efficiently alkylated with a set of alcohols (**201**) under Fukuyama modification of the Mitsunobu reaction. Resin **202** underwent sulfonamide removal, followed by acylation with substituted pent-4-enoic acids to give precursor **204** which, under RCM conditions with 5 mol% of first-generation Grubbs catalyst (**192**), afforded the library of Freidinger ε-lactams **205** in acceptable overall yields (Scheme 49).

Oxygen-containing heterocycles have been also obtaining via RCM on solid phase. The described strategy was based on the use of D-(+)-mannitol as a chiral template for the synthesis of *cis*-fused pyranofurans properly functionalized for conformationally constrained peptidomimetic precursors (Scheme 50) [154]. Immobilized tetrahydrofuran **206** was derived from D-(+)-mannitol and tethered to Rink amide resin using BOP [benzotriazol-l-yl-oxy-tris-(dimethylamino) phosphoniumhexafluorophosphate] in the presence of DIPEA. Then, functionalization of the hydroxyl group was achieved using different iso-cyanates to give carbamates, followed by reduction of the azide and *N*-acylation, to provide the corresponding immobilized diolefins **207**. Finally, cyclization/cleavage

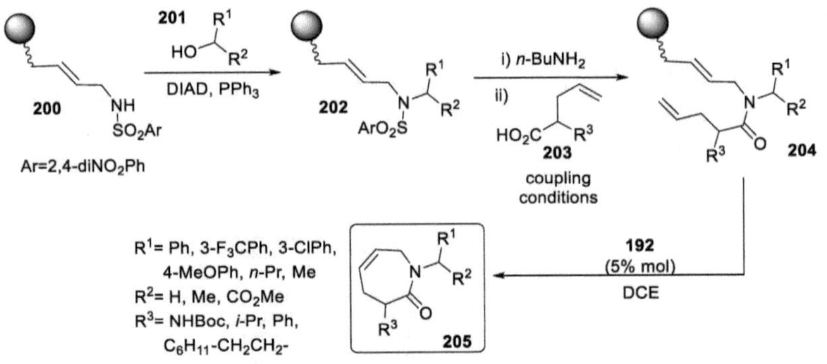

Scheme 49 Solid-phase synthesis of Freidinger ε-lactams

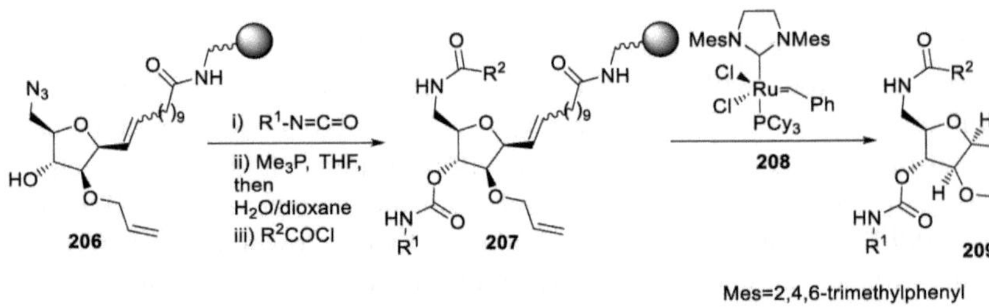

Scheme 50 Synthesis of pyranofurans by cleavage/cyclization using RCM

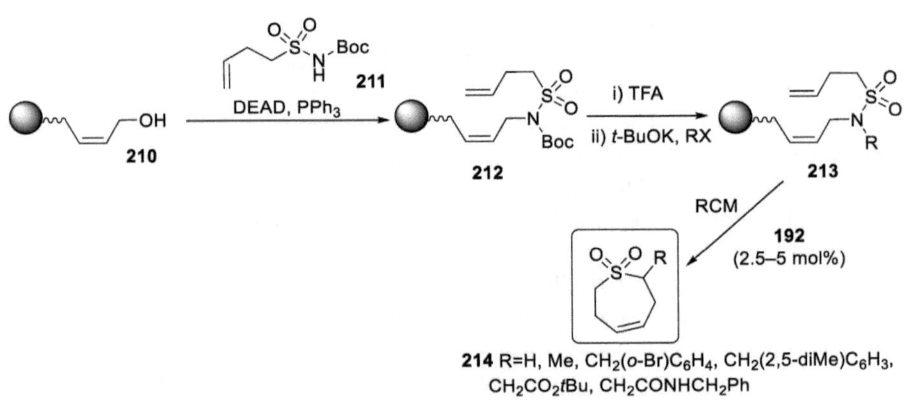

Scheme 51 Solid-phase RCM for the synthesis of *N*-alkylated cyclic sulfonamides

was carried out by using 5 mol% of second-generation Grubbs precatalyst (**208**) to afford the pyranofurans **209** in good to excellent yields after flash column chromatography.

Brown et al. [155] have also used the cleavage/cyclization approach to the synthesis of seven-membered cyclic sulfonamides (Scheme 51). Allylic alcohol tethered to Merrifield resin (**210**), used as starting material, underwent Mitsunobu

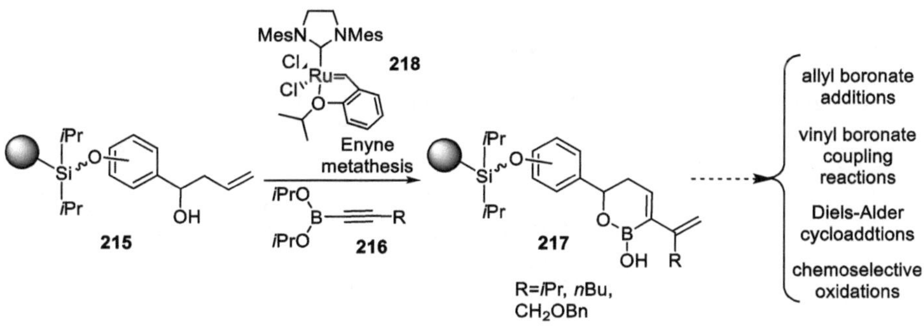

Scheme 52 Enyne metathesis for the synthesis of immobilized cyclic dialkenylboronic acids

coupling with sulfonamide **211** to give the diene **212**. In order to incorporate diversity, *N*-alkylation was carried out to obtain the immobilized cyclization precursors **213**. Finally, RCM was performed using first-generation Grubbs precatalyst (**192**) to afford a small library of the corresponding *N*-alkylated cyclic sulfonamides **214**.

A comprehensive analysis was developed by the authors to optimize reaction conditions. They remarked the necessity of column chromatography purification due to the presence of ruthenium-derived impurities. Besides, addition of an olefin to the RCM step proved to be deleterious for the synthetic process: no better yield was obtained and purity of the cyclic sulfonamides decreased significantly due to cross-metathesis by-products.

Schreiber et al. [156] have reported a study towards the synthesis of immobilized cyclic dialkenylboronic acids (**217**) by an enyne coupling (Scheme 52). These cyclic dialkenylboronic acids were planning to be used as common intermediate for generating skeletal diversity. Methodology was based on the reaction of the homoallylic alcohol **215** tethered to a diisopropylsilyl ether linker [157] with alkynyl boronates **216** under Ru-carbene catalysis. The process can be considered as an intramolecular enyne metathesis since an initial transesterification is performed to bind alkyne and alkene components together, in order to facilitate metathesis conditions. Best results were obtained with the "phosphine-free" Hoveyda–Grubbs precatalyst (**218**) [158].

In a very interesting approach, Heerding and coworkers [159] have developed the combinatorial synthesis of tricyclic heterocycles by a tandem intramolecular enyne metathesis/Diels–Alder cycloaddition (Scheme 53). The Wang resin-bound 4-hydroxybenzoic acid (**219**) was converted to the corresponding allyl amide **221**, and alkyne insertion was performed by *t*BuOLi and an alkynyl mesylate to yield the enyne structure **223**. The RCEM was carried out with first-generation Grubbs catalyst (**192**) and followed by the Diels–Adler reaction with maleimide to afford the hexahydroisoindoles **225**. Varying R^1 and R^2 substituents, the dienophile, and the aromatic ring, a library of about 4,000 compounds was generated by this synthetic sequence.

Apart from metathesis reactions, ruthenium has been another catalysis used by Deiters et al. to obtain biologically attractive structures by solid-phase [2 + 2 + 2]

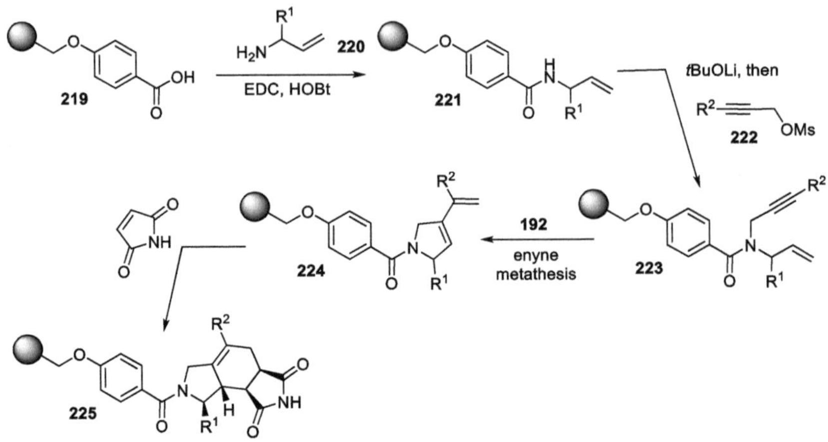

Scheme 53 Solid-phase tandem intramolecular enyne metathesis/Diels–Alder cycloaddition

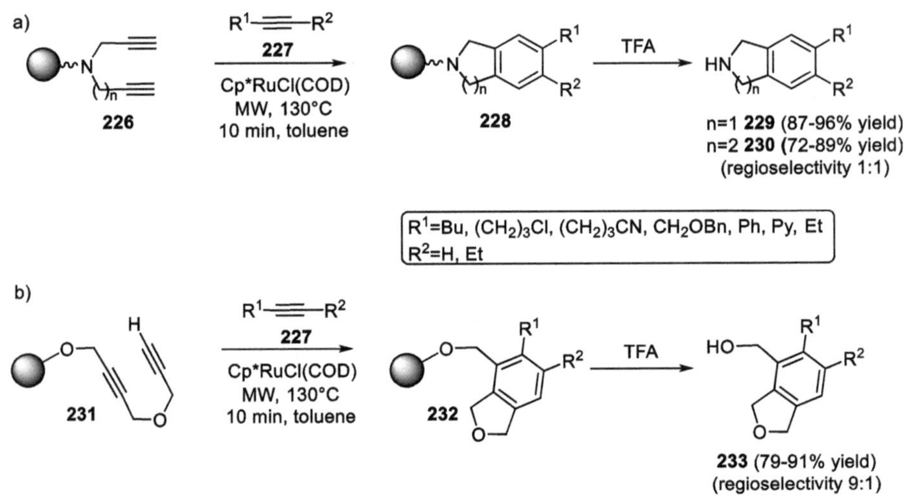

Scheme 54 Isoindolines, tetrahydroisoquinolines, and phthalans by solid-phase [2 + 2 + 2] cyclotrimerization under Ru catalysis

cyclotrimerization (Scheme 54) [160]. Trityl resin-immobilized diynes **226** reacted efficiently with a series of substituted alkynes **227** in the presence of Cp*RuCl (COD) [Cp* = η^5-C_5Me_5, COD = 1,5-cyclooctadiene] as catalyst [161], under microwave heating. Cleavage from the resin with TFA afforded the isoindolines **229** and tetrahydroisoquinolines **230** (1:1 mixture of regioisomers) in high yields (Scheme 54a). As already mentioned, this solid-phase [2 + 2 + 2] cyclotrimerization avoids oligomerization of the starting immobilized diyne. Combination of Ru catalysis and microwave irradiation was an important improvement comparing with previous results using Wilkinson's catalyst under thermal conditions (see

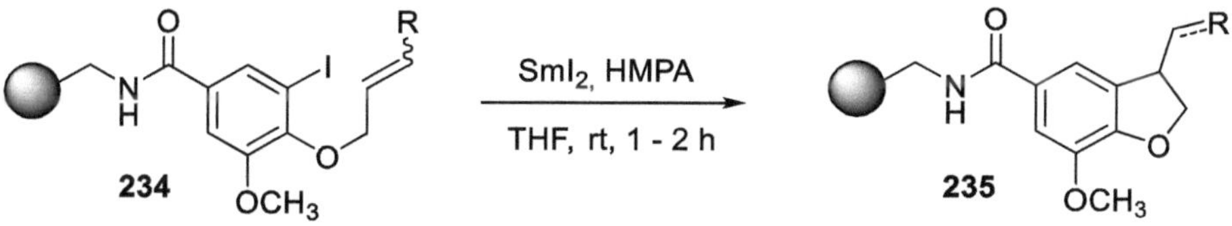

Scheme 55 Synthesis of benzofuran derivatives **235** via key SmI_2-mediated radical cyclization

Sect. 2.8, Scheme 46) [138]. In the case of oxygen-containing immobilized diynes, attachment to resin was performed by one of the sides of the moiety, such as diyne **231** (Scheme 54b). Again, optimized conditions gave good isolated yields of the corresponding phthalans **233**, after cleavage from the resin. Authors reported that a carboxyl linkage worked well for the cyclotrimerization under conventional heating [138] although, for the sterically demanding trityl linker, microwave irradiation was more efficient.

2.10 *Samarium*

Samarium diiodide (SmI_2) is one of the most important reducing agents available to the synthetic organic chemist. In addition, SmI_2 can mediate a wide variety of organic chemistry reactions, which generally do not possess an alternative reagent system [162]. Actually, SmI_2 is often highly chemo- and stereoselective. Interestingly, this metal can participate in both radicals and anionic processes. Furthermore, its reactivity can be modulated by the addition of various salts and cosolvents to the reaction mixture. After a long initial role in inorganic chemistry, since 1977, SmI_2 has been introduced to the synthetic chemistry community to enable unexpected chemical transformations. Undoubtedly, the use of this reagent on solid-phase synthesis has been focused mainly as highly selective cleavage conditions for specific linkers [163], such as α-Hetero-Atom Substituted Carbonyl (HASC) traceless linkers studied by Procter [164–167], and hydroxylamine-based linkers (traceless cleavages of N–O bond linkage) [168, 169]. Further applications, and in particular towards supported cyclizations, are more limited. So far the reported examples consist in the synthesis of various benzofuran derivatives (**235**) from alkenyl-substituted aryl iodides (**234**) through SmI_2-mediated radical cyclizations (Scheme 55) [170, 171]. The annulation proceeded in mild and simple conditions, improving previous procedures. In these cases, the SmI_2 was used in excess (10 equiv.).

2.11 *Tin*

Tin has been known since time immemorial, and the discovery of its alloy with copper started the Bronze Age in about 3,500 BC. The most common tin ore is

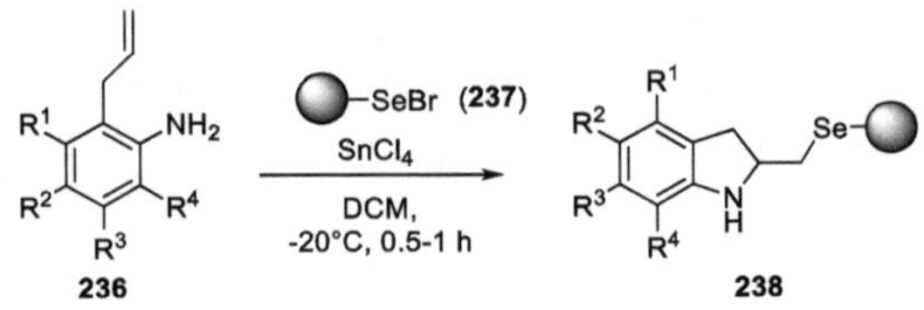

Scheme 56 Cycloloading mediated by $SnCl_4$ to afford immobilized indolines on selenyl resin

cassiterite, SnO_2. As tin industrial applications are broad (solders, tin plating, especially alloys), the extraction of this metal is active and continuous as well as its recovery from the commercial tin containers, thus this metal can be easily supplied. Regarding organotin compounds, their first application consisted in the stabilization of PVC against heating process in about 1943 [172]. Since then, during the last half of twentieth century and so on, there has been a broad research in organotin chemistry. The main drawback related to the organotin compounds is the difficulty for their quantitative removal from the reaction mixture, which is especially unfavorable due to tin toxicity. This disadvantage limits considerably the tin exploitation as reagent and/or catalyst in biomedical and pharmaceutical synthesis (both in classical solution conditions and SPOS) [173]. Concerning the use of tin compounds in cyclization on solid support, the examples consisted mainly in the coupling of *o*-substituted *o*-allyl anilines (**236**) onto a polystyrene-based selenyl bromide resin (**237**) via a 5-*exo*-trig cyclization to afford resin-bound indoline scaffolds **238** with good loadings (Scheme 56) [174, 175]. The cycloloading reaction proceeded smoothly in a short period of time (0.5–1 h) and low temperature (−20°C) and only required the presence of the $SnCl_4$ as Lewis acid. This cycloloading allowed the traceless synthesis of diverse polycyclic substituted indolines.

2.12 *Titanium*

Titanium is, after iron, the most abundant transition metal on earth being, in addition, this metal and its derivatives nontoxic and inexpensive compounds. Considering these benefits, the synthesis of various heterocyclic compounds such as pyrazoles, pyrimidines, isoxazoles, quinolines, pyrroles, hydrazones, and indoles using catalysts based on titanium has been reported [176]. Furthermore, various methodologies of titanium-based catalysis involve the application of multicomponent reactions in the generation of heterocycles [177].

In an already mentioned article, Bräse describe the synthesis of phthalides under a variety of conditions, including a titanium tetrachloride-mediated cyclization/ cleavage approach (Scheme 57) [91]. In this case, using a Sakurai-type addition, Merrifield resin-immobilized *ortho*-formyl benzoic acids **239** were treated with 5 equiv. of both allyltrimethyl silanes **240** and $TiCl_4$, to afford the corresponding phthalides **241a** ($R^1 = H$) in fairly good yields and purity. For $R^1 = OMe$, yields of phthalides **241b** were mostly lower.

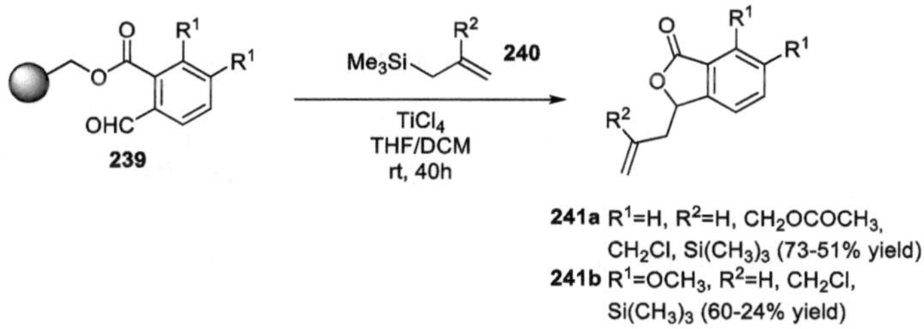

Scheme 57 Titanium tetrachloride-mediated solid-phase synthesis of phthalides

2.13 Zinc

Zinc is an interesting choice for organometallic chemistry since it presents a high abundance in the earth's crust, has biological relevance, and possesses a series of chemical abilities. Many important applications of organozinc compounds in organic chemistry have been known and commonly applied for long time, such as in breakthrough procedures like Reformatsky reaction, Frankland–Duppa reaction, Negishi reaction, and Fukuyama reaction, among others [178]. However, in comparison to other metals, the interest in zinc as catalyst core has been underdeveloped. Recently, this situation changed along with the search of "greener" chemistry; thus more attention has been placed on the use of this abundant metal with low price and low toxicity [179]. Actually, in the last three decades an increasing number of publications regarding zinc as reagent and/or catalyst have appeared and the number of achievements applying zinc in organic transformations continues to rise [180]. In particular, the use of Zn for the promotion of cyclizations on insoluble polymer has been implemented towards two different heterocyclic scaffolds. The first report in this area consisted in the preparation of diverse immobilized 2-arylindols scaffolds (**244**) from commercially available building blocks (Scheme 58) [181]. For this purpose, a flexible solid-phase protocol based on Fisher indol synthesis enabled the production of the desired substituted indoles (**244**) from the supported ketones **242** and phenylhydrazine (**243**), in the presence of $ZnCl_2$, under acidic conditions (4-hydroxymethylbenzoic acid (HMB) linker).

An interesting cyclization mediated by Zn(II) consists in the cycloaddition of dienes and double bonds leading to 6-membered rings. The first application of this methodology involved iterative Diels–Alder reactions between different keto-conjugated alkenes and diverse dienes in the presence of $ZnCl_2$ leading to carbonated polycyclic products [182]. Taking advantage of this strategy, Waldman et al. treated Wang or HMBA resin-supported tryptophan imines (**245**) with electron-rich dienes (**246**) to yield enaminones (**247**) via Zn-mediated tandem Mannich–Michael reactions (Scheme 59) [183, 184]. Based on a biology-oriented synthesis (BIOS), the immobilized heterocycles **247** were subjected to further modifications to produce four natural product-inspired medium-sized heterocycle collections.

As previously described, Zn(II) has been also utilized for the synthesis of triazole rings [34] and benzofuranone skeletons [91].

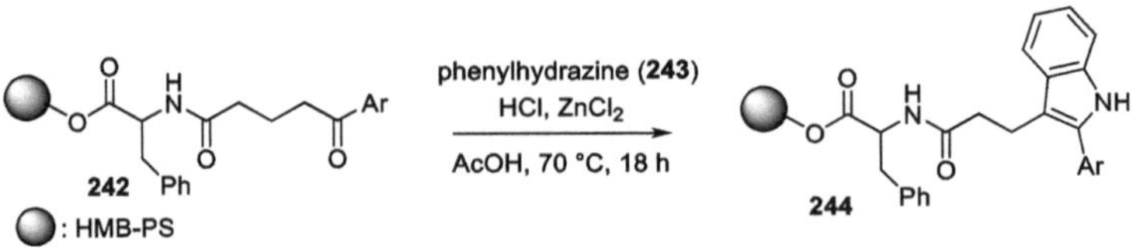

Scheme 58 Fisher indol synthesis of immobilized 2-arylindols (**244**)

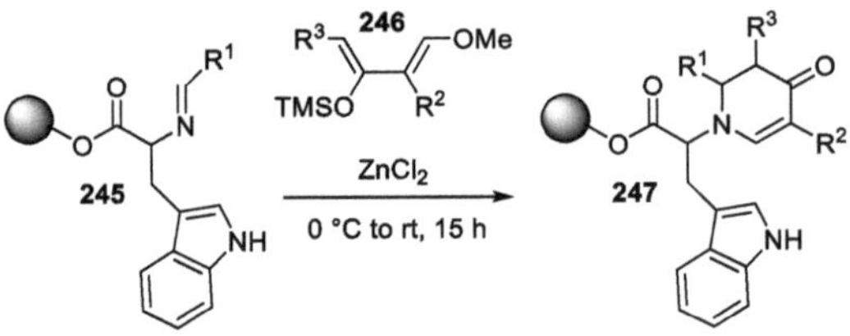

Scheme 59 Zn-mediated tandem Mannich–Michael reactions towards immobilized enaminones **247**

3 Conclusions

This chapter is focused on the previously unreviewed application of metal-mediated solid-phase chemistry to the synthesis of heterocycles. Examples demonstrate the usefulness of the combination of SPOS with organometallic chemistry to the generation of libraries of heterocyclic structures. Some revival has been noticed in solid-phase chemistry when new (and old) comparative advantages started to be recognized and exploited. Apart from the classical advantages concerning to an easy purification and automatization for generating libraries, solid-phase synthesis offers an important contribution to green chemistry by reducing solvent waste due to less chromatographic separations comparing with traditional organic synthesis. In the case of transition metal-catalyzed cross-coupling reactions, an increase in chemo- and regioselectivity can be achieved. While excess of the non-immobilized substrate can be used to take the reaction to reach completion, homodimerization of the immobilized substrate is less favorable due to the separation between the reactive sites at the resin.

We think that this chapter could be very helpful for perceiving the possibilities of applying the solid-phase version of metal-mediated reactions to the preparation of biologically promising heterocyclic compounds.

References

1. Nadin A, Hattotuwagama C, Churcher I (2012) Angew Chem Int Ed 51:1114
2. MacLellan P, Nelson A (2013) Chem Commun 49:2383
3. Hartwig JF (2010) Organotransition metal chemistry: from bonding to catalysis. University Science Books, New York

4. Galloway WRJD, Isidro-Llobet A, Spring DR (2010) Nat Commun 1:80
5. O'Connor CJ, Laraia L, Spring DR (2011) Chem Soc Rev 40:4332
6. O'Connor CJ, Beckmann HSG, Spring DR (2012) Chem Soc Rev 41:4444
7. Testero SA, Mata EG (2008) J Comb Chem 10:487
8. Yellol GS, Sun CM (2012) Solid-supported synthesis. In: Green techniques for organic synthesis and medicinal chemistry. Wiley, Chichester, p 393
9. Joule JA, Mills K (2010) Heterocyclic chemistry, 5th edn. Wiley-Blackwell, Hoboken
10. Gosmini C, Moncomble A (2010) Isr J Chem 50:568
11. Tilly D, Dayaker G, Bachu P (2014) Catal Sci Technol 4:2756
12. Senaiar RS, Young DD, Deiters A (2006) Chem Commun 2006:1313
13. Young D, Deiters A (2007) Angew Chem Int Ed 46:5187
14. Hassan J, Sevignon M, Gozzi C, Schulz E, Lemaire M (2002) Chem Rev 102:1359
15. Attanasi OA, De Crescentini L, Filippone P, Mantellini F, Tietze LF (2001) Tetrahedron 57:5855
16. Masse CE, Ng PY, Fukase Y, Sanchez-Rosello M, Shaw JT (2006) J Comb Chem 8:293
17. Dai WM, Guo DS, Sun LP, Huang XH (2003) Org Lett 5:2919
18. Sun LP, Dai WM (2006) Angew Chem Int Ed 45:7255
19. Kolb HC, Sharpless KB (2003) Drug Discovery Today 8:1128
20. Kolb HC, Finn MG, Sharpless KB (2001) Angew Chem Int Ed 40:2004
21. Castro V, Rodriguez H, Albericio F (2016) ACS Comb Sci 18:1
22. Tornoe CW, Meldal M (2001) Peptidotriazoles: copper(I)-catalyzed 1,3-dipolar cycloadditions on solid-phase. In: Lebl M, Houghten RA (eds) Peptides: the wave of the future. Springer, San Diego, p 263
23. Tornoe CW, Christensen C, Meldal M (2002) J Org Chem 67:3057
24. Meldal M, Tornoe CW, Nielsen TE, Diness F, Le Quement ST, Christensen CA, Jensen JF, Worm-Leonhard K, Groth T, Bouakaz L, Wu B, Hagel G, Keinicke L (2010) Biopolymers 94:161
25. Tornoe CW, Sanderson SJ, Mottram JC, Coombs GH, Meldal M (2004) J Comb Chem 6:312
26. Vilá S, Badosa E, Montesinos E, Feliu L, Planas M (2015) Eur J Org Chem 2015:1117
27. Fittler H, Avrutina O, Glotzbach B, Empting M, Kolmar H (2013) Org Biomol Chem 11:1848
28. Gopi HN, Tirupula KC, Baxter S, Ajith S, Chaiken IM (2006) Chem Med Chem 1:54
29. Franke R, Doll C, Eichler J (2005) Tetrahedron Lett 46:4479
30. Turner RA, Oliver AG, Lokey RS (2007) Org Lett 9:5011
31. Jagasia R, Holub JM, Bollinger M, Kirshenbaum K, Finn MG (2009) J Org Chem 74:2964
32. Punna S, Kuzelka J, Wang Q, Finn MG (2005) Angew Chem Int Ed 44:2215
33. Angelo NG, Arora PS (2005) J Am Chem Soc 127:17134
34. Angelo NG, Arora PS (2007) J Org Chem 72:7963
35. Proteau-Gage A, Rochon K, Roy M, Albert PJ, Guerin B, Gendron L, Dory YL (2013) Bioorg Med Chem Lett 23:5267
36. Iwata T, Tanaka K, Tahara T, Nozaki S, Onoe H, Watanabe Y, Fukase K (2013) Mol BioSyst 9:1019
37. Norgren AS, Budke C, Majer Z, Heggemann C, Koop T, Sewald N (2009) Synthesis 2009:488
38. St. Amant AH, Engbers C, Hudson RHE (2013) Artif DNA PNA XNA 4:4
39. Gasser G, Husken N, Koster SD, Metzler-Nolte N (2008) Chem Commun 2008:3675
40. Husken N, Gasser G, Koster SD, Metzler-Nolte N (2009) Bioconjugate Chem 20:1578
41. Patra M, Metzler-Nolte N (2011) Chem Commun 47:11444
42. Gasser G, Jager K, Zenker M, Bergmann R, Steinbach J, Stephan H, Metzler-Nolte N (2010) J Inorg Biochem 104:1133
43. Gasser G, Pinto A, Neumann S, Sosniak AM, Seitz M, Merz K, Heumann R, Metzler-Nolte N (2012) Dalton Trans 41:2304
44. Mansawat W, Boonlua C, Siriwong K, Vilaivan T (2012) Tetrahedron 68:3988
45. Liu Y, Wang XF, Chen Y, Zhang LH, Yang ZJ (2012) Med Chem Commun 3:506

46. Abramova T (2013) Molecules 18:1
47. Amblard F, Cho JH, Schinazi RF (2009) Chem Rev 109:4207
48. Qiu J, El-Sagheer AH, Brown T (2013) Chem Commun 49:6959
49. Paritala H, Suzuki Y, Carroll KS (2013) Tetrahedron Lett 54:1869
50. Bouillon C, Meyer A, Vidal S, Jochum A, Chevolot Y, Cloarec JP, Praly JP, Vasseur JJ, Morvan F (2006) J Org Chem 71:4700
51. Pourceau G, Meyer A, Vasseur JJ, Morvan F (2009) J Org Chem 74:1218
52. Astakhova IK, Wengel J (2013) Chem Eur J 19:1112
53. Cornier PG, Boggián DB, Mata EG, Delpiccolo CML (2012) Tetrahedron Lett 53:632
54. Cornier PG, Delpiccolo CML, Mascali FC, Boggian DB, Mata EG, Cardenas MG, Blank VC, Roguin LP (2014) Med Chem Commun 5:214
55. Oh S, Jang HJ, Ko SK, Ko Y, Park SB (2010) J Comb Chem 12:548
56. Goess BC, Hannoush RN, Chan LK, Kirchhausen T, Shair MD (2006) J Am Chem Soc 128:5391
57. Castro V, Rodriguez H, Albericio F (2013) Org Lett 15:246
58. Rodriguez Loaiza P, Lober S, Hubner H, Gmeiner P (2007) Bioorg Med Chem 15:7248
59. Lober S, Rodriguez-Loaiza P, Gmeiner P (2003) Org Lett 5:1753
60. Lober S, Gmeiner P (2004) Tetrahedron 60:8699
61. Leeb L, Gmeiner P, Lober S (2007) QSAR Comb Sci 26:1145
62. Bettinetti L, Lober S, Hubner H, Gmeiner P (2005) J Comb Chem 7:309
63. Rodriguez Loaiza P, Lober S, Hubner H, Gmeiner P (2006) J Comb Chem 8:252
64. Rodriguez Loaiza P, Lober S, Hubner H, Gmeiner P (2009) Bioorg Med Chem 17:5482
65. Papadopoulos A, Shiao TC, Roy R (2012) Mol Pharm 9:394
66. Li Y, Zhang B, Hoskins JN, Grayson SM (2012) J Polym Sci A Polym Chem 50:1086
67. Guo Z, Lei A, Liang X, Xu Q (2006) Chem Commun 2006:4512
68. Hashmi AS (2016) Chem Rev 107:3180
69. Arcadi A (2008) Chem Rev 108:3266
70. Das A, Sohel SM, Liu RS (2010) Org Biomol Chem 8:960
71. Furstner A, Davies PW (2007) Angew Chem Int Ed 46:3410
72. Furstner A (2009) Chem Soc Rev 38:3208
73. Hashmi AS, Hutchings GJ (2006) Angew Chem Int Ed 45:7896
74. Hashmi AS (2014) Acc Chem Res 47:864
75. Li Z, Brouwer C, He C (2008) Chem Rev 108:3239
76. Rudolph M, Hashmi AS (2012) Chem Soc Rev 41:2448
77. Lein M, Rudolph M, Hashmi SK, Schwerdtfeger P (2010) Organometallics 29:2206
78. Pernpointner M, Hashmi AS (2009) J Chem Theory Comput 5:2717
79. Hashmi AS (2003) Gold Bull 36:3
80. Corma A, Leyva-Perez A, Sabater MJ (2011) Chem Rev 111:1657
81. Hashmi ASK (2010) Angew Chem Int Ed 49:5232
82. La-Venia A, Testero SA, Mischne MP, Mata EG (2012) Org Biomol Chem 10:2514
83. La-Venia A, Medran NS, Krchnak V, Testero SA (2016) ACS Comb Sci 18:482
84. Bauer I, Knolker HJ (2008) Iron complexes in organic chemistry. In: Plietker B (ed) Iron catalysis in organic chemistry: reactions and applications. Wiley-VCH Verlag GmbH & Co. KGaA, Weinheim, p 1
85. Bauer I, Knolker HJ (2015) Chem Rev 115:3170
86. Bolm C, Legros J, Le Paih J, Zani L (2004) Chem Rev 104:6217
87. Kilburn JP, Lau J, Jones RCF (2003) Tetrahedron Lett 44:7825
88. Severinsen R, Kilburn JP, Lau JF (2005) Tetrahedron 61:5565
89. Yamamoto H, Oshima K (2006) Main group metals in organic synthesis. John Wiley-VCH Verlag GmbH & Co. KGaA, Weinheim
90. Knepper K, Brase S (2003) Org Lett 5:2829
91. Knepper K, Ziegert RE, Brase S (2004) Tetrahedron 60:8591
92. Chen C, Li X, Neumann CS, Lo MMC, Schreiber SL (2005) Angew Chem Int Ed 44:2249

93. Sebahar PR, Williams RM (2000) J Am Chem Soc 122:5666
94. Abel EW, MacKillop A, Stone FGA, Wilkinson G (1995) Main-group metal organometallics in organic synthesis. Elsevier Science & Technology Books, Exeter
95. Larock R (2012) Organomercury compounds in organic synthesis. Springer, Berlin
96. Boeglin D, Cantel S, Heitz A, Martinez J, Fehrentz JA (2003) Org Lett 5:4465
97. Hitotsuyanagi Y, Motegi S, Fukaya H, Takeya K (2002) J Org Chem 67:3266
98. Johansson Seechurn CCC, Kitching MO, Colacot TJ, Snieckus V (2012) Angew Chem Int Ed 51:5062
99. Tsuji J (2006) Palladium reagents and catalysts: new perspectives for the 21st century. Wiley, Chichester
100. Selander N, Szabo J (2011) Chem Rev 111:2048
101. Brown RCD (1998) J Chem Soc Perkin Trans 1:3293
102. Goff DA, Zuckermann RN (1995) J Org Chem 60:5748
103. Wang Y, Huang TN (1998) Tetrahedron Lett 39:9605
104. Berteina S, Wendeborn S, De Mesmaeker A (1998) Synlett 1998:1231
105. Bolton GL, Hodges JC (1999) J Comb Chem 1:130
106. Gracias V, Moore JD, Djuric SW (2004) Tetrahedron Lett 45:417
107. Chen W, Li Z, Ou L, Giulianott MA, Houghten RA, Yu Y (2011) Tetrahedron Lett 52:1456
108. Dadiboyena S, Nefzi A (2012) Tetrahedron Lett 53:6897
109. Fancelli D, Fagnola MC, Severino D, Bedeschi A (1997) Tetrahedron Lett 38:2311
110. Hu Y, Nawoschik KJ, Liao Y, Ma J, Fathi R, Yang Z (2004) J Org Chem 69:2235
111. Cao J, Huang X (2010) J Comb Chem 12:1
112. Zhang HC, Maryanoff BE (1997) J Org Chem 62:1804
113. Grigg R, Major JP, Martin FM, Whittaker M (1999) Tetrahedron Lett 40:7709
114. Brown RCD, Fisher M (1999) Chem Commun 1999:1547
115. Ma S, Duan D, Shi Z (2000) Org Lett 2:1419
116. Akaji K, Kiso Y (1997) Tetrahedron Lett 38:5185
117. Collini MD, Ellingboe JW (1997) Tetrahedron Lett 38:7963
118. Fagnola MC, Candiani I, Visentin G, Cabri W, Zarini F, Mongelli N, Bedeschi A (1997) Tetrahedron Lett 38:2307
119. Wu TYH, Ding S, Gray NS, Schultz PG (2001) Org Lett 3:3827
120. Grigg R, MacLachlan WS, MacPherson DT, Sridharan V, Suganthan S (2001) Tetrahedron 57:10335
121. Yu Y, Ostresh JM, Houghten RA (2003) Tetrahedron Lett 44:2569
122. Yamazaki K, Kondo Y (2002) Chem Commun 2002:210
123. Yamazaki K, Kondo Y (2002) J Comb Chem 4:191
124. Yamazaki K, Nakamura Y, Kondo Y (2003) J Org Chem 68:6011
125. Yun W, Mohan R (1996) Tetrahedron Lett 37:7189
126. Arumugam V, Routledge A, Abell C, Balasubramanian S (1997) Tetrahedron Lett 38:6473
127. Zhang HC, Brumfield KK, Maryanoff BE (1997) Tetrahedron Lett 38:2439
128. Zhang HC, Brumfield KK, Jaroskova L, Maryanoff BE (1998) Tetrahedron Lett 39:4449
129. Zhang HC, Ye H, Moretto AF, Brumfield KK, Maryanoff BE (2000) Org Lett 2:89
130. Zhang HC, Ye H, White KB, Maryanoff BE (2001) Tetrahedron Lett 42:4751
131. Evans PA (2005) Modern rhodium-catalyzed organic reactions. Wiley-VCH Verlag GmbH & Co. KGaA, Weinheim
132. Gowravaram MR, Gallop MA (1997) Tetrahedron Lett 38:6973
133. Whitehouse DL, Nelson J, Savinov SN, Austin DJ (1997) Tetrahedron Lett 38:7139
134. Whitehouse DL, Nelson KH, Savinov J, Lowe RS, Austin DJ (1998) Bioorg Med Chem 6:1273
135. Savinov SN, Austin DJ (2002) Org Lett 4:1419
136. Iso Y, Shindo H, Hamana H (2000) Tetrahedron 56:5353
137. Sun Q, Zhou X, Islam K, Kyle DJ (2001) Tetrahedron Lett 42:6495
138. Young DD, Senaiar RS, Deiters A (2006) Chem Eur J 12:5563

139. Murahashi SI (2004) Ruthenium in organic synthesis. Wiley-VCH Verlag GmbH & Co. KGaA, Weinheim
140. Dias EL, Nguyen ST, Grubbs RH (1997) J Am Chem Soc 119:3887
141. Astruc D (2005) New J Chem 29:42
142. Poulsen CS, Madsen R (2003) Synthesis 1:0001
143. Kotha S, Meshram M, Tiwari A (2009) Chem Soc Rev 38:2065
144. Deiters A, Martin SF (2004) Chem Rev 104:2199
145. Schuster M, Pernerstorfer J, Blechert S (1996) Angew Chem Int Ed Engl 35:1979
146. Nguyen ST, Grubbs RH, Ziller JW (1993) J Am Chem Soc 115:9858
147. Grubbs RH, Chang S (1998) Tetrahedron 54:4413
148. Schrodi Y, Pederson RL (2007) Aldrichim Acta 40:45
149. van Maarseveen JH, den Hartog JAJ, Engelen V, Finner E, Visser G, Kruse CG (1996) Tetrahedron Lett 37:8249
150. Veerman JJN, van Maarseveen JH, Visser GM, Kruse CG, Schoemaker HE, Hiemstra H, Rutjes FPJT (1998) Eur J Org Chem 1998:2583
151. Piscopio AD, Miller JF, Koch K (1997) Tetrahedron Lett 38:7143
152. Piscopio AD, Miller JF, Koch K (1998) Tetrahedron Lett 39:2667
153. Perdih A, Kikelj D (2006) Curr Med Chem 13:1525
154. Timmer MSM, Verdoes M, Sliedregt LAJM, van der Marel GA, van Boom JH, Overkleeft HS (2003) J Org Chem 68:9406
155. Moriggi JD, Brown LJ, Castro JL, Brown RCD (2004) Org Biomol Chem 2:835
156. Brittain DEA, Gray BL, Schreiber SL (2005) Chem Eur J 11:5086
157. Hu Y, John A, Labadie JW, Gooding OW, Trost BM (1998) J Org Chem 63:4518
158. Kingsbury JS, Harrity JPA, Bonitatebus PJ, Hoveyda AH (1999) J Am Chem Soc 121:791
159. Heerding DA, Takata DT, Kwon C, Huffman WF, Samanen J (1998) Tetrahedron Lett 39:6815
160. Young DD, Sripada L, Deiters A (2007) J Comb Chem 9:735
161. Yamamoto Y, Arakawa T, Ogawa R, Itoh K (2003) J Am Chem Soc 125:12143
162. Procter DJ, Flowers RA, Skrydstrup T (2009) Organic synthesis using samarium diiodide. A practical guide. The Royal Society of Chemistry, Cambridge
163. Mirizzi D, Pulici M (2011) Molecules 16:3252
164. McAllister LA, Brand S, de Gentile R, Procter DJ (2003) Chem Commun 2003:2380
165. McAllister LA, Turner KL, Brand S, Stefaniak M, Procter DJ (2006) J Org Chem 71:6497
166. McKerlie F, Rudkin IM, Wynne G, Procter DJ (2005) Org Biomol Chem 3:2805
167. Turner KL, Baker TM, Islam S, Procter DJ, Stefaniak M (2006) Org Lett 8:329
168. Meloni MM, Taddei M (2001) Org Lett 3:337
169. Myers RM, Langston SP, Conway SP, Abell C (2000) Org Lett 2:1349
170. Du X, Armstrong RW (1997) J Org Chem 62:5678
171. Du X, Armstrong RW (1998) Tetrahedron Lett 39:2281
172. Davies AG (2008) Introduction and overview. Tin chemistry. John Wiley & Sons, Ltd, Chichester, p. 1
173. Otera J, Biesemans M, Pinoie V, Poelmans K, Willem R, Blank WJ, Hessell ET, Kinart WJ, Kinart CM (2008) Tin in catalysis. Tin chemistry. John Wiley & Sons, Ltd, Chichester, p. 667
174. Nicolaou KC, Roecker AJ, Pfefferkorn JA, Cao GQ (2000) J Am Chem Soc 122:2966
175. Nicolaou KC, Roecker AJ, Hughes R, van Summeren R, Pfefferkorn JA, Winssinger N (2003) Bioorg Med Chem 11:465
176. Nakamura I, Yamamoto Y (2004) Chem Rev 104:2127
177. Odom AL, McDaniel TJ (2015) Acc Chem Res 48:2822
178. Wu XF, Neumann H (2012) Adv Synth Catal 354:3141
179. Enthaler S (2013) ACS Catal 3:150
180. Enthaler S, Wu XF (2015) Zinc catalysis: applications in organic synthesis. Wiley-VCH Verlag GmbH & Co.KGaA, Weinheim
181. Hutchins SM, Chapman KT (1996) Tetrahedron Lett 37:4869

182. Winkler JD, Kwak YS (1998) J Org Chem 63:8634
183. Corrêa Jr IR, Nören-Müller A, Ambrosi HD, Jakupovic S, Saxena K, Schwalbe H, Kaiser M, Waldmann H (2007) Chem Asian J 2:1109
184. Nören-Müller A, Corrêa Jr IR, Prinz H, Rosenbaum C, Saxena K, Schwalbe HJ, Vestweber D, Cagna G, Schunk S, Schwarz O, Schiewe H, Waldmann H (2006) Proc Natl Acad Sci U S A 103:10606

Top Heterocycl Chem (2017) 52: 139–166
DOI: 10.1007/7081_2016_5

Published online: 13 April 2017

Intramolecular Arylation of 2-Nitrobenzenesulfonamides: A Route to Diverse Nitrogenous Heterocycles

Eva Schütznerová and Viktor Krchňák

Abstract In addition to being an integral component of numerous drugs, 2- and 4-nitrobenzenesulfonyl (Nos) groups are used as protecting/activating agents for the selective alkylation of primary amines. 2-Nos amides are also useful intermediates for the synthesis of fused nitrogenous heterocycles. This chapter focuses on the use of 2-Nos amides as intramolecular arylation agents and on the application of *C*- and *N*-aryl derivatives for the synthesis of diverse heterocycles in the solid phase.

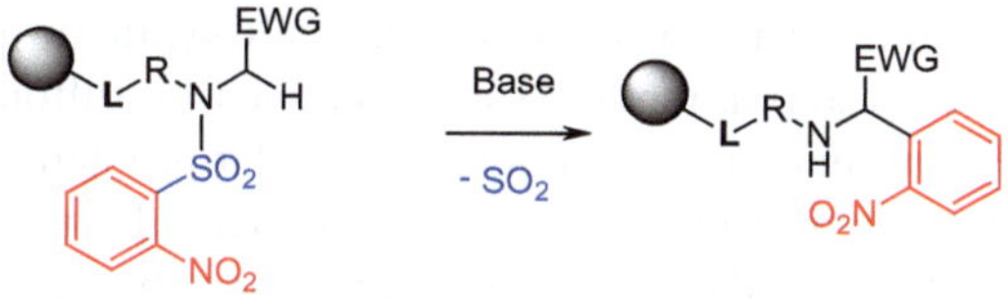

Keywords Arylation • Heterocycles • Nitrobenzenesulfonamides • Solid-phase synthesis

E. Schütznerová
Faculty of Medicine, Institute of Molecular and Translation Medicine, Palacký University, Hněvotínská 5, 779 00 Olomouc, Czech Republic

V. Krchňák (✉)
Department of Organic Chemistry, Faculty of Science, Palacký University, Trida 17, Listopadu 12, 771 46 Olomouc, Czech Republic

Department of Chemistry and Biochemistry, University of Notre Dame, 251 Nieuwland Science Center, Notre Dame, IN 46556, USA
e-mail: vkrchnak@nd.edu

Contents

1 Introduction

Fukuyama et al. reported in 1995 that 2-nitrobenzenesulfonyl (Nos) and 4-Nos amides are excellent activating/protecting groups for the regioselective *N*-alkylation of primary amines to their corresponding secondary amines [1]. At that time, selective monoalkylation was considered a difficult synthetic problem. Conventional methods such as alkylation with alkyl halides or sulfonates can result in tertiary amines or quaternary ammonium salts. Similar problems occur with reductive alkylation. A general solution to the challenge of monoalkylation is the Fukuyama modification of the Mitsunobu alkylation by reacting 2- or 4-Nos derivatives (**1**, Scheme 1) with alcohols. The protecting Nos group can then be easily removed through a nucleophilic aromatic substitution mechanism using a mercaptoethanol/DBU cleavage cocktail via formation of the Meisenheimer complex (**4**) to afford the desired secondary amine (**5**).

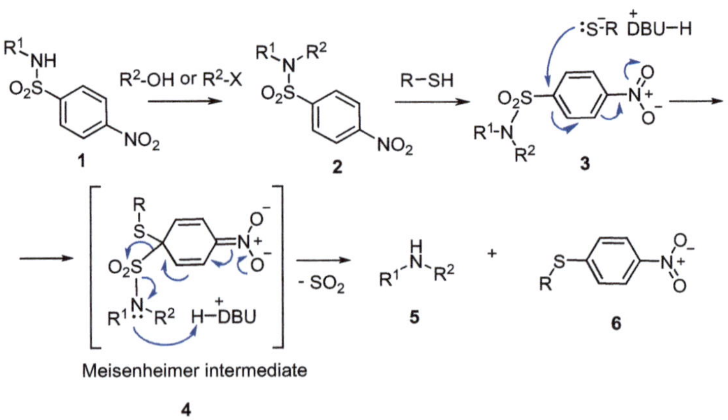

Scheme 1 Fukuyama alkylation using the 4-Nos activating group

This methodology quickly attracted much attention and has become very popular in solid-phase synthesis, initially in peptide chemistry [2, 3] and subsequently in the synthesis of polyfunctional nonpeptidyl molecules, for example, in the solid-phase synthesis of heterocycles such as 4,7,8-trisubstituted 1,2,3,4-tetrahydrobenzo [*e*][1,4] diazepin-5-ones [4] and trisubstituted benzo[1,4]diazepin-5-ones [5]. The applications of Nos amides in synthetic solid-phase organic chemistry are detailed in an excellent review by Fülöpová and Soural [6].

2 Nos Amides in Heterocyclic Chemistry

In addition to the use of the Nos group as an activating/protecting group for the alkylation of primary amines, Nos amides have been advantageously incorporated into reaction products. Three different scenarios of Nos incorporation have been developed: (1) integration of an intact Nos group as part of the target structure (Fig. 1, structure **I**); (2) reduction of the nitro group and integration into a heterocycle (structure **II**); and (3) the use of Nos as a source of *C*/*N*-arylation via formation of a C–C or C–N bond (structure **III**). Figure 1 shows the fragment included in the final structure in red. In this chapter, the second and third scenarios are discussed.

3 Heterocycles via Intramolecular Arylation

3.1 Nos Amides as Intramolecular Arylation Agents

The mechanism of Nos group removal, briefly discussed in the previous section, is accomplished by nucleophilic attack of sulfur on the aromatic carbon creating (in this case) an unstable Meisenheimer intermediate (**4**, Scheme 1). However, nucleophilic attack by an internal nucleophile can occur, as first documented by Klebs (Scheme 2), where the Nos group undergoes a Smiles-type rearrangement [7]. Klebs observed transformation of *N*-(2-hydroxyethyl)-4-nitrobenzenesulfonamide (**7**) to *N*-(2-hydroxyethyl)-4-nitroaniline (**10**) via a tandem Smiles rearrangement. The result of this rearrangement was the *N*-aryl derivative and concurrent release of SO_2.

An analogous C–C bond formation was also described in the base-mediated intramolecular nucleophilic aromatic substitution reaction to form methyl-

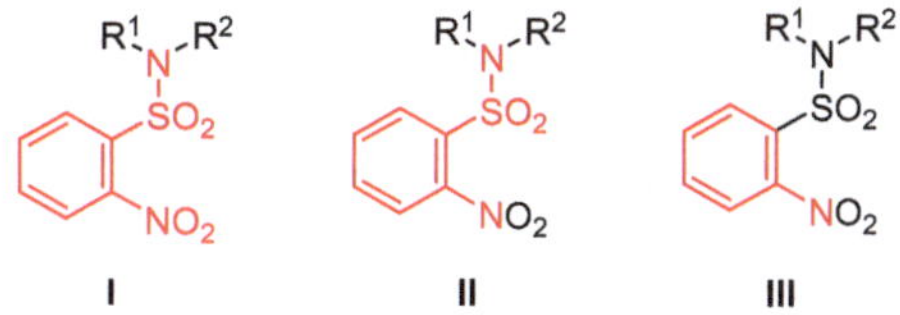

Fig. 1 Three scenarios for integration of the Nos group into products

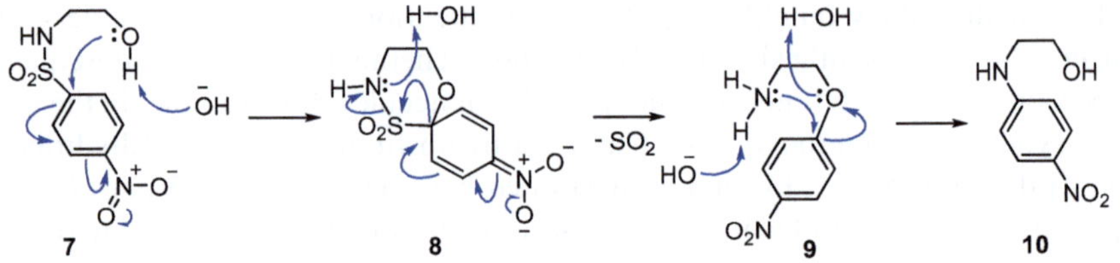

Scheme 2 Intramolecular *N*-arylation via a tandem Smiles rearrangement

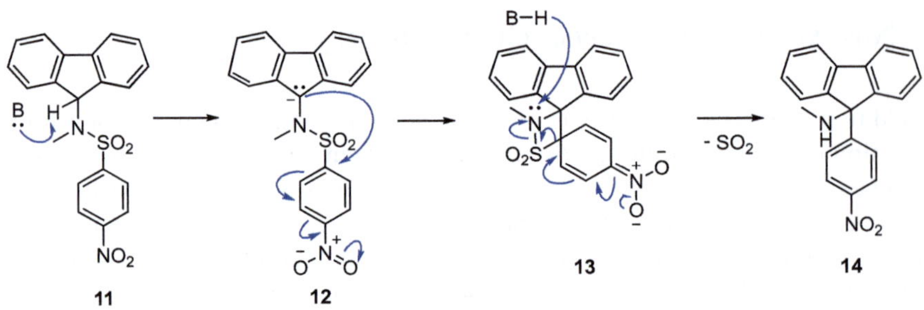

Scheme 3 Base-catalyzed reaction of the 4-Nos derivative via an intramolecular aromatic S_N reaction

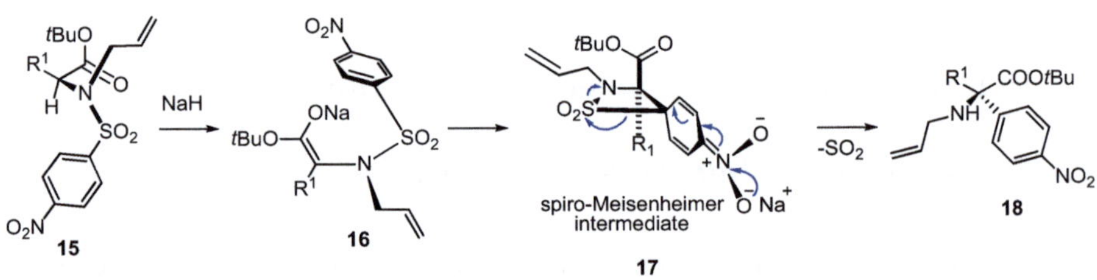

Scheme 4 Proposed mechanism of sulfonamide rearrangement via a spiro-Meisenheimer intermediate

[9-(4-nitrophenyl)-9*H*-fluoren-9-yl]-amine (**14**, Scheme 3) from the corresponding 4-Nos amide (**11**) [8].

The formation of an interesting by-product from the alkylation during the Fukuyama procedure using both protecting groups (2- and 4-Nos) was first reported by Bowman [9]. However, the authors did not report the complete structural characterization of this impurity. The *C*-arylated structure was elucidated by Wilson [10] in 1999, and a mechanism for this sulfonamide rearrangement was proposed to proceed via a Meisenheimer intermediate. The stereochemistry was further investigated by Lupi (Scheme 4) [11], who identified the intermediate as a spiro-Meisenheimer intermediate (**17**).

The nucleophilic attack by an internal *C*-nucleophile on the solid phase was also realized. 2-Nos amides (**19**, Scheme 5), prepared by alkylation using bromoketones, underwent a DBU-catalyzed C–C bond formation with the concurrent release of SO_2. The mechanism is analogous to the previous scheme and proceeded via the spiro-Meisenheimer intermediate (**20**) [12].

This reaction appeared to be general for 2-Nos amides containing an acidic proton at the alpha carbon and was achieved by the presence of an electron-withdrawing substituent (CO-R^3 in Scheme 5) at the C^{α} carbon. Scheme 6 represents the generic structure of the substrates that undergo base-mediated C–C bond formation. The presence of the nitro group in the ortho position provided numerous opportunities to use the *C*-aryl derivative for the synthesis of diverse nitrogenous heterocycles. Scheme 6 shows six structural types that underwent the intramolecular arylation. In the following sections, we discuss the variability of heterocycles accessible from the *C*-aryl derivatives.

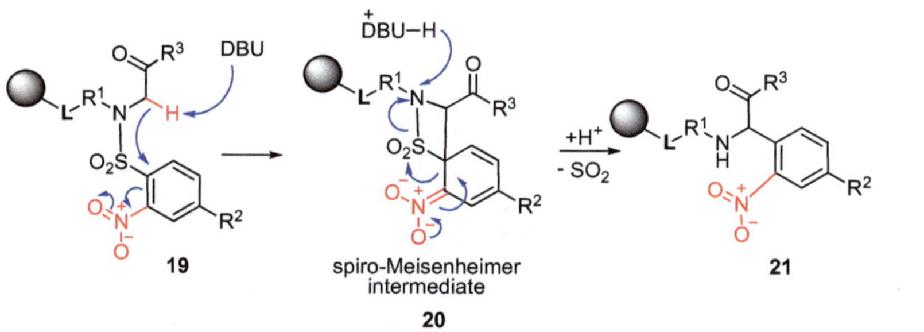

L-R¹ = Rink-NHCO-$(CH_2)_{2-3}$-, Rink-NHCO-$CH(CH_3)$-, Wang-OCO-$(CH_2)_2$-, Wang-O-$(CH_2)_2$-, Wang-OCO-NH-$(CH_2)_3$-
R² = H, 4-NO_2, 4-NH_2, 4-OMe, 4-CF_3
R³ = -OEt, 4-OMe-Ph, 4-Me-Ph, 4-Cl-Ph, 4-NH_2-3,5-diCl-Ph, 3-NO_2-Ph

Scheme 5 Intramolecular *C*-arylation on the solid phase

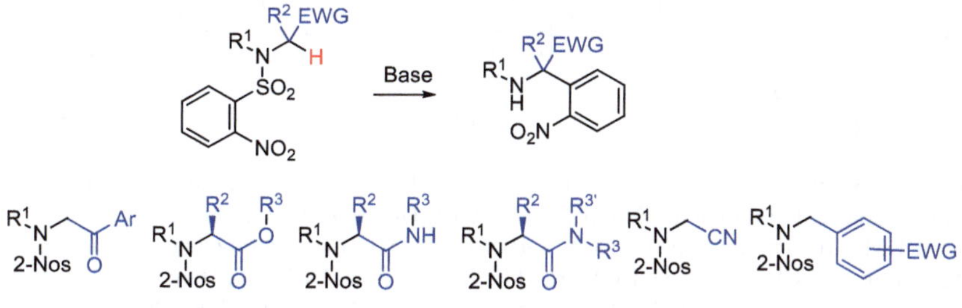

Scheme 6 Generic structures amenable to *C*-arylation

3.2 Common Route to the Solid-Phase Synthesis of Nos Intermediates

An important aspect of our solid-phase synthesis of heterocycles via arylation chemistry is that acyclic intermediates for all diverse heterocycles were prepared using the same synthetic route. This route included only three steps while incorporating three diversity positions using commercially available building blocks under mild conditions (Scheme 7). Synthesis of model compounds for subsequent arylation started with preparation of resin-bound primary amines (**22a–d**). *N*-Fmoc amino acids, *N*-Fmoc-amino alcohols and diamines were immobilized onto an acid-labile Wang resin. Rink amide resin was also acylated with Fmoc amino acids. Removal of the Fmoc protecting group was followed by sulfonylation with 2-Nos-Cl's and *N*-alkylation using alcohols via the Fukuyama-Mitsunobu protocol or electrophiles (typically a bromoketone or α-bromo ester). These resin-bound compounds (**24**) contained three diversity positions and were termed advanced intermediates.

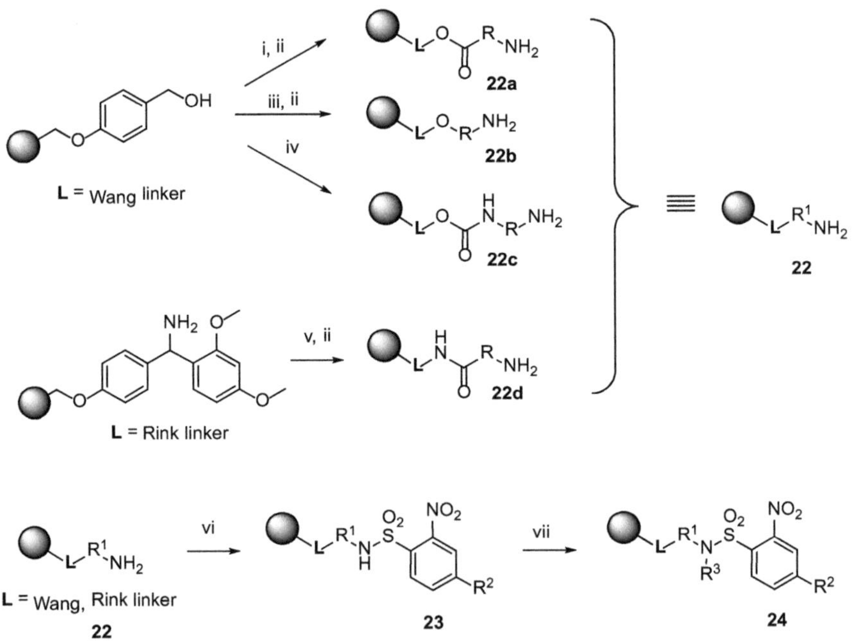

Scheme 7 Synthesis of Nos intermediates amenable to arylation. Reagents and conditions: (i) Fmoc-protected amino acid, DMAP, DIC, HOBt, DCM/DMF (1:1), rt., 16 h; (ii) 50% piperidine in DMF, rt., 15 min; (iii) CCl_3CN, DBU, DCM, 1 h, then Fmoc-ethanolamine, $BF_3{\cdot}Et_2O$, THF, 30 min; (iv) CDI, pyridine, DCM, rt., 3 h, then diamine, DCM, rt., 16 h; (v) Fmoc-protected amino acid, DIC, HOBt, DCM/DMF (1:1), rt., 16 h; (vi) 2-Nos-Cl, lutidine, DCM, rt., 16 h; (vii) bromoketones, DIEA, DMF, rt., 16 h or alcohols, PPh_3, DIAD, THF, rt., overnight

3.3 2-Nos Amino Ketones

Resin-bound acyclic intermediates prepared by alkylation of 2-Nos amides with bromoketones provided a route to several different heterocycles. Scheme 8 summarizes individual paths to target compounds. Each route is then discussed individually.

3.3.1 Indazole Oxides and Indazoles

Indazole oxide was the first heterocycle synthesized using the Nos arylation strategy; Bouillon et al. reported a tandem C–C formation reaction followed by N–N bond formation leading to indazole oxides (**27**, Scheme 9) [12]. The initially formed *C*-aryl derivative (**21**) reacted to form a five-membered ring by N–N bond formation. This unstable intermediate (**25**) dehydrated to yield indazole oxides (**26**). In the second part of this tandem reaction, the N–N bond formed through a

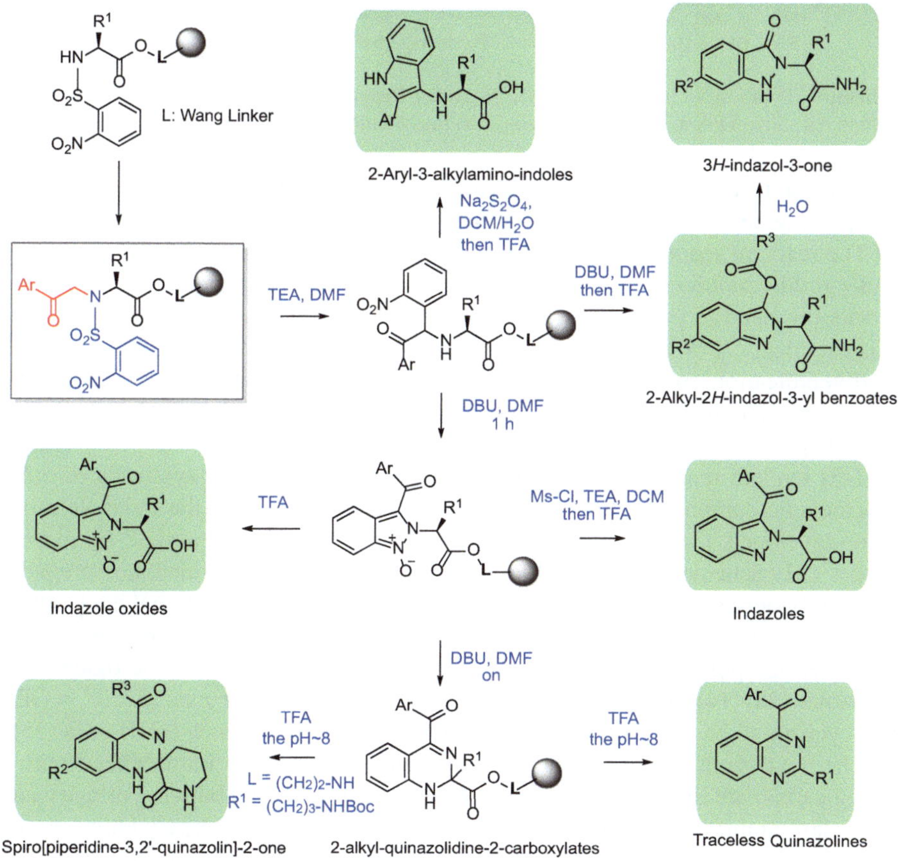

Scheme 8 Overview of heterocycles prepared from 2-Nos amino ketones

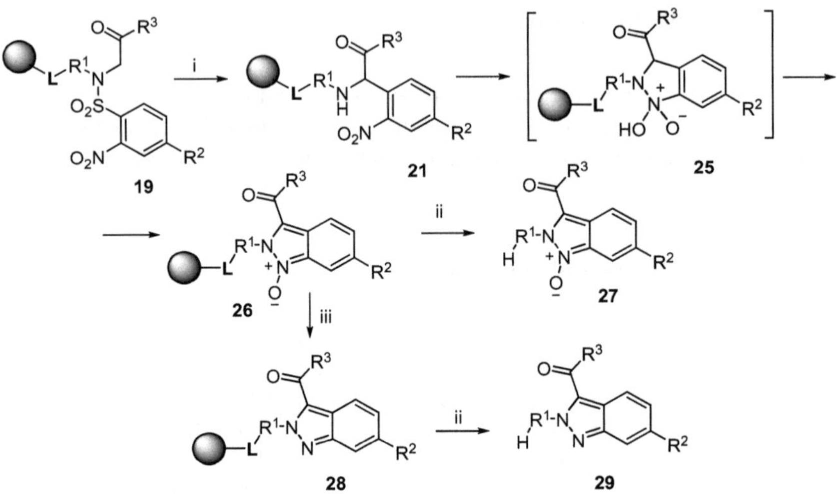

L-R¹ = Rink-NHCO-$(CH_2)_{2\text{-}3}$-, Rink-NHCO-CH(CH_3)-, Wang-OCO-$(CH_2)_2$-, Wang-O-$(CH_2)_2$-, Wang-OCO-NH-$(CH_2)_3$-

R² = -H, -NO_2, -NH_2, -OMe, -CF_3

R³ = -OEt, 4-OMe-Ph, 4-Me-Ph, 4-Cl-Ph, 4-NH_2-3,5-diCl-Ph, 3-NO_2-Ph

Scheme 9 Synthesis of indazoles and their *N*-oxides. Reagents and conditions: (i) DBU, DMF, rt., 30 min; (ii) 50% TFA in DCM, rt., 30 min; (iii) mesyl chloride, TEA, DCM, rt., 16 h

condensation reaction between the nitro and amino groups; such bond formation had already been observed in the synthesis of 3,4-fused cinnolines in 1960 [13].

The reaction sequence was remarkably efficient. Indazole oxides were prepared under mild conditions (room temperature) using commercially available building blocks (no need to prepare dedicated synthons), and the purity of crude products was excellent (72–99%). The synthesis provided indazole oxides with three points of diversification and tolerated a wide range of substituents. Deoxygenation by mesyl chloride in the presence of a base yielded the corresponding indazoles (**28**) [12].

This tandem reaction offered numerous possibilities for the synthesis of other heterocyclic compounds. First, diverse compounds comprising the indazole motif were developed. Compounds with additional fused rings, pyrazino[1,2-*b*]indazole oxides (**32**, Scheme 10), were obtained from resin-bound intermediates prepared using polymer-supported ethylenediamine via cyclization of an internal nucleophile (the amino group of ethylenediamine) with a carbonyl group. The authors further expanded the portfolio of target compounds by reducing the indazole oxides into indazoles (**33**) [14]. Alkylation with bromoacetate yielded a precursor (**33**, R^3 = OEt) that cyclized in basic media to 3,4-dihydropyrazino[1,2-*b*]indazole-1(2H)-ones (**35**).

The next synthesis was carried out on diethylenetriamine immobilized onto a Wang carbamate linker. Reaction with 2-Nos-Cl sulfonylated both the primary and secondary amino groups (**36**, Scheme 11). The Nos at the secondary amino group served as a protecting group, and the Nos primary amino group was activated

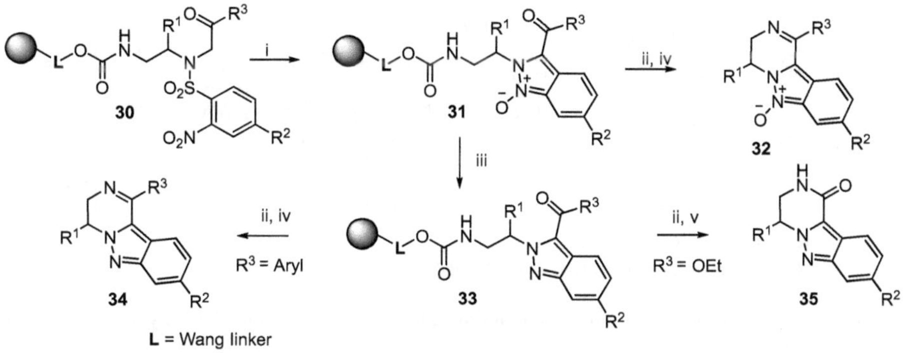

Scheme 10 Synthesis of 3,4-dihydropyrazino[1,2-*b*]indazoles and their *N*-oxides. Reagents and conditions: (i) DBU, DMF, rt., 30 min; (ii) 50% TFA in DCM, rt., 1 h; (iii) mesyl chloride, TEA, DCM, rt., 16 h; (iv) AcOH, rt., 2 h; (v), TEA, MeOH, rt., 16 h

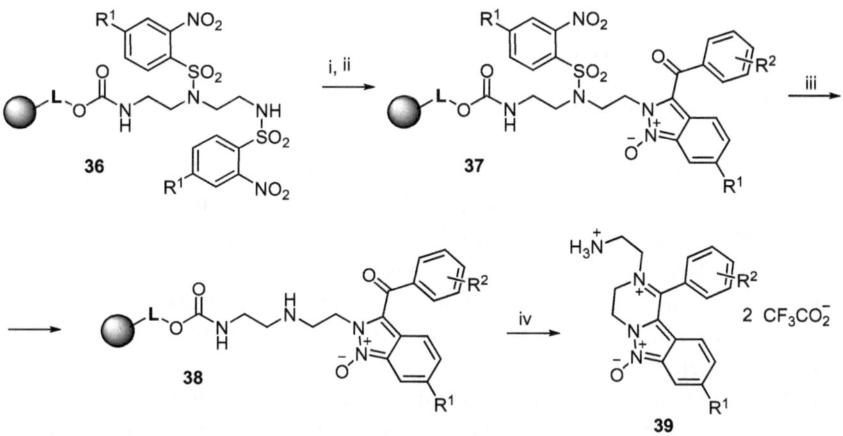

Scheme 11 Solid-phase synthesis of 2-(2-amino/hydroxyethyl)-1-aryl-3,4-dihydropyrazino[1,2-*b*]indazol-2-iums. Reagents and conditions: (i) bromoketones, DIEA, DMF, rt., 16 h; (ii) DBU, DMF, rt., 30 min; (iii) 2-mercaptoethanol, DBU, DMF, rt., 5 min; (iv) 50% TFA in DCM, rt., 1 h

for subsequent alkylation with bromoketones. After base-mediated formation of indazole oxides (**37**), the Nos group was cleaved and the resin-bound construct (**38**) was exposed to the cleavage cocktail. The secondary amino group enabled the formation of iminium ions via condensation with carbonyls that were stable as TFA salts (**39**) [15]. The authors also reported deoxygenation leading to the indazole derivatives.

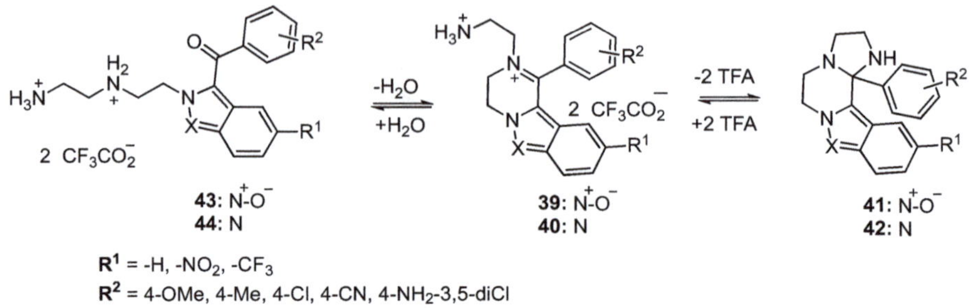

Scheme 12 Equilibrium iminium-imidazolidine and rearrangement into 2,3-dihydro-1*H*-imidazo [1,2-*b*]indazoles

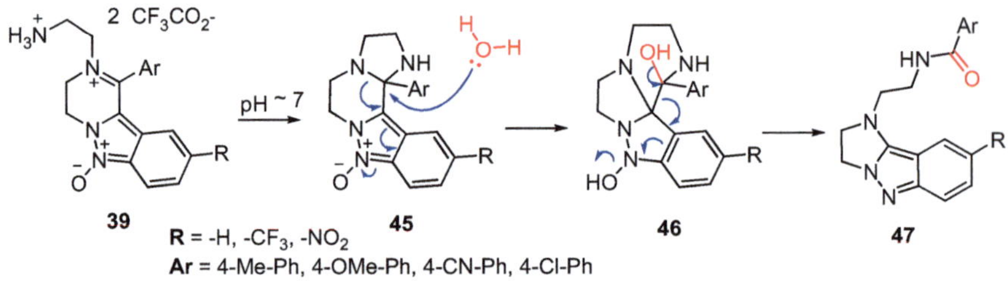

Scheme 13 Mechanism of rearrangement to 2,3-dihydro-1*H*-imidazo[1,2-*b*]indazoles

TFA-mediated cleavage from the resin yielded iminium trifluoroacetes (**39**, **40**, Scheme 12) [15]. The presence of a nucleophile on the side-chain of the amino group enabled ring closure at neutral pH and the formation of complex fused heterocycles, derivatives of imidazolidines (**41**, **42**). By contrast, the iminiums exhibited a tendency to hydrolyze in aqueous solutions into keto-amines (**43**, **44**). The relative ratio of this equilibrium was dependent on the substitution pattern (R^1 and R^2 substituents).

Further extension of the iminium indazole chemistry represented an unprecedented rearrangement of fused indazole oxides (**39**, Scheme 13) at neutral pH to 2,3-dihydro-1*H*-imidazo[1,2-*b*]indazoles (**47**) [16]. The rearrangement formally involves a concomitant 5-membered ring-opening, 6- to 5-membered ring contraction, amide formation, and deoxygenation.

Another synthetically useful application of the arylation chemistry was discovered during a detailed study of the indazole oxide formation mechanism. The tandem C–C followed by N–N bond formation initially formed an unstable nonplanar five-membered intermediate (**25**, Scheme 14, also Scheme 9), which spontaneously dehydrated to the indazole oxide (**26**, Scheme 9). Although this intermediate (**25**) could not be isolated, the reaction conditions under which the cyclic intermediate underwent an acid-mediated intramolecular Baeyer-Villiger oxidation were identified, where the *N*-oxide precursor served as an intramolecular oxidizing agent. The solid-phase synthesis played a critical role here because fast isolation of the intermediate by washing of the resin enabled rapid access to the unstable Baeyer-Villiger oxidation-prone intermediate. In addition, the cleavage

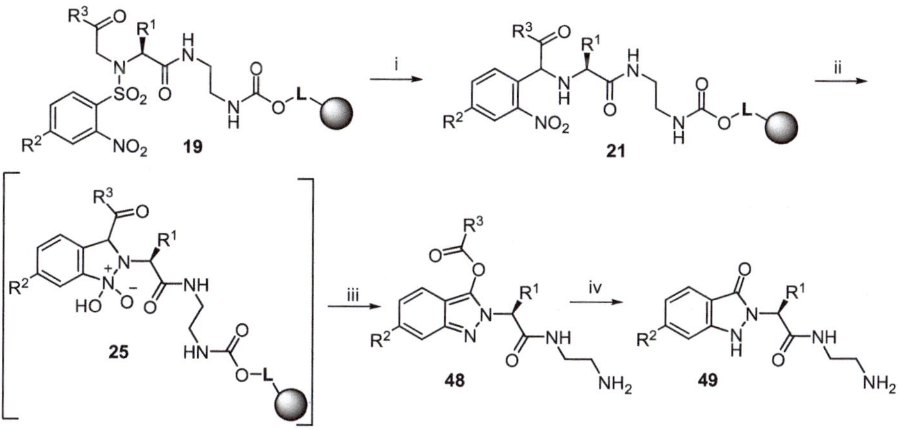

L = Wang linker
R¹ = -Me, -CH$_2$O-*t*Bu, -(CH$_2$)$_4$-NH$_2$, -Ph
R² = -H, -OMe, -CF$_3$
R³ = 4-OMe-Ph, 4-Cl-Ph, Ph

Scheme 14 Esters from ketones: the synthesis of 2-alkyl-2*H*-indazol-3-yl benzoates and 2-alkyl-1,2-dihydro-3*H*-indazol-3-ones. Reagents and conditions: (i) 0.5 M DIEA, DMF, rt., 16 h; (ii) 0.2 M DBU, DMF, rt., 30 min; (iii) 50% redistilled TFA, anhydrous DCM, rt., 1 h; (iv) H_2O, MeOH, rt., overnight

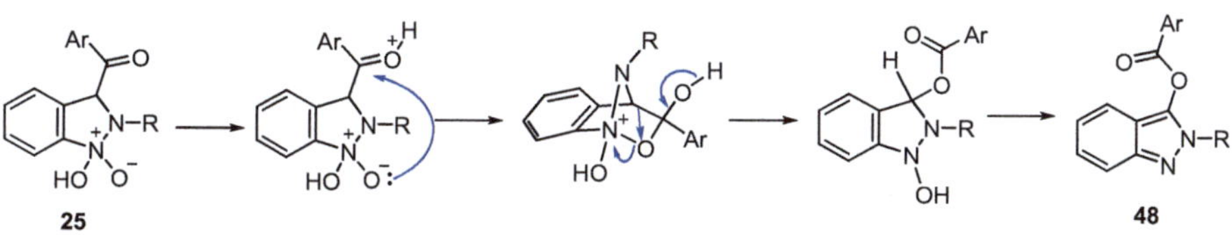

Scheme 15 Mechanism of ketone oxidation by the internal *N*-oxide

conditions were critical for performing the Baeyer-Villiger oxidation leading to the indazole ester. The purity of the TFA used in the cleavage cocktail strongly affected the reaction outcome. Indazole oxide was formed exclusively with the lower-quality synthesis-grade TFA (purity <99%); reagent-grade TFA (99%) led to a mixture, and clean conversion to the indazole ester was observed only with redistilled TFA (purity ≥99%).

This synthetic route provided access to 2-alkyl-2*H*-indazol-3-yl benzoates (**48**, Scheme 14). Hydrolysis of the ester (**48**) yielded pharmacologically relevant 2-alkyl-1,2-dihydro-3*H*-indazol-3-ones (**49**) [17]. Notably, the typical external oxidant for Baeyer-Villiger syntheses, 3-chloroperbenzoic acid, reportedly did not provide the expected product (only traces were detected).

The proposed mechanism of the Baeyer-Villiger oxidation using the *N*-oxide as an intramolecular oxidant (Scheme 15) was enabled by the nonplanar structure of the unstable intermediate (**25**).

3.3.2 Quinazolines

Indazoles and related compounds were not the only family of heterocycles accessible from the Nos intermediates. While further extending the scope of the indazole oxide chemistry, Krupkova et al. [18] observed that the indazole oxide (**26**, Scheme 16) synthesized using glycine as the first building block underwent DBU-mediated ring expansion to quinazolines (**50**). The key structural feature was the presence of an acidic alpha proton that triggered the ring-expansion and formation of the quinazoline. Expanding indazole oxides to quinazolines is an example of scaffold hopping [19, 20], a concept used in drug development based on a modification of the central core of the molecule, wherein the peripheral substituents remain unchanged.

The role of the acidic alpha proton is illustrated in the mechanism of quinazoline formation (**50**, Scheme 17) [18]. No precedent for the transformation of indazole oxides to quinazolines was found, only a distant analogy of the ring expansion of indazoles [21].

A drawback of the synthesis is the presence of a carboxamide in all the synthesized compounds; thus, the synthesis is not traceless because all compounds are amides of quinazoline-2-carboxylic acids. Fulopova et al. reported an extension of *C*-arylation methodology covering traceless synthesis, where the target compounds (**54**, Scheme 18) have no "trace" of the linker used for immobilization of the first building block [22]. They carried out the synthesis using α-amino acids. After

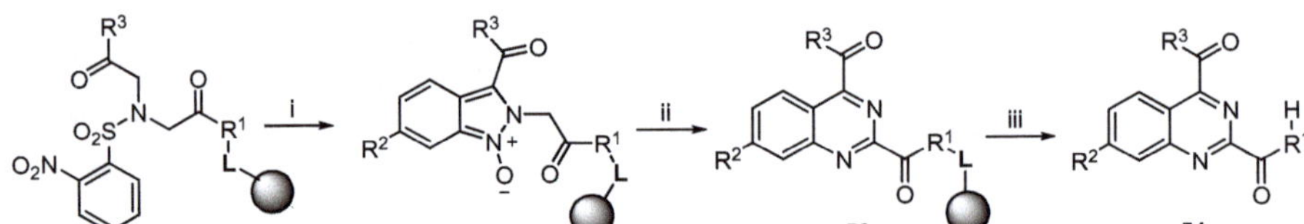

R¹ = NH; **L** = Rink amide linker
R¹ = NH-alkyl; **L** = BAL linker
R² = -H, -NO₂, -CF₃
R³ = -OEt, -Ph, 4-OMe, 4-Me, 4-Cl, 4-CN, 4-NH₂-3,5-diCl-Ph, piperidin-1-yl, 4-Ph-piperazin-1-yl

Scheme 16 Synthesis of quinazolines. Reagents and conditions: (i) DBU, DMF, rt., 30 min; (ii) DBU, DMF, rt., 10 min to 16 h; (iii) 50% TFA in DCM, rt., 1 h

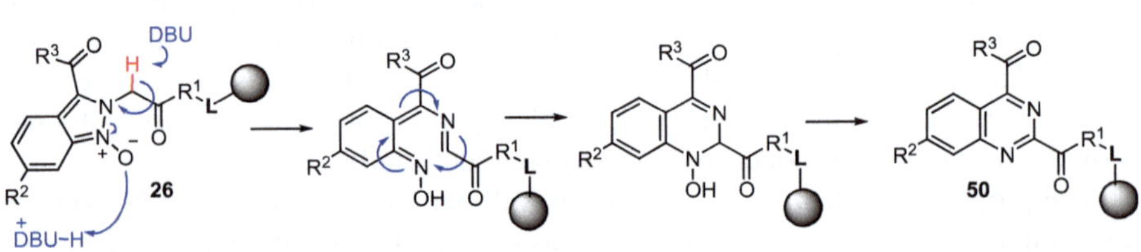

Scheme 17 Mechanism of ring expansion of indazole oxides to quinazolines

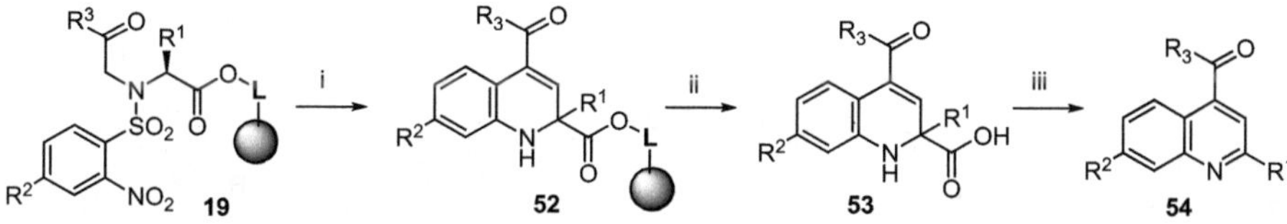

L = Wang linker
R^1 = -Me, $-CH_2-OH$, $-(CH_2)_4-NH_2$, $-CH(CH_3)_2$, $-CH_2-Ph$, $-(CH_2)_2COOH$
R^2 = -H, $-NO_2$, $-CF_3$, -OMe
R^3 = -Ph, 4-OMe-Ph, 4-Me-Ph, $4-CF_3-Ph$, 4-Cl-Ph, 4-CN-Ph

Scheme 18 Synthesis of trisubstituted quinazolines. Reagents and conditions: (i) DBU, DMF, rt., 30 min to 16 h; (ii) 50% TFA in DCM, rt., 1 h; (iii) neutralization with ammonium acetate, purification by C18 cartridge

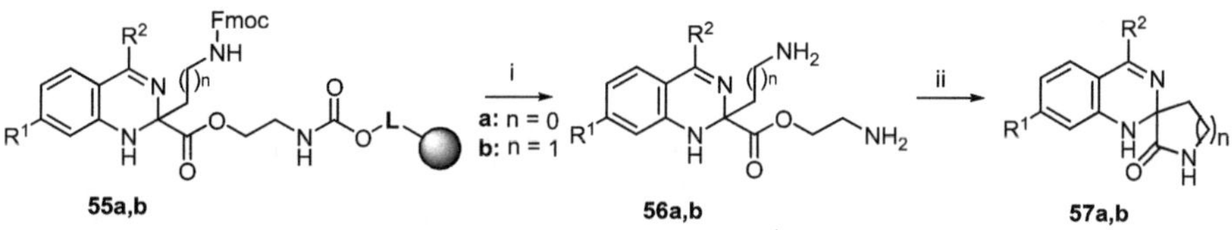

L = Wang linker
R^1 = -H, -Cl, $-CF_3$, -OMe
R^2 = -Bz, 4-MeO-Bz, $4-CF_3-Bz$

Scheme 19 Synthesis of quinazoline spiro derivatives. Reagents and conditions: (i) 50% TFA, DCM, rt., 1 h; (ii) TEA, DMF, rt., 5 min

cleavage from the resin, the carboxylate compounds (**53**) spontaneously decarboxylated in aqueous ammonium acetate/acetonitrile solution at ambient temperature to yield the aromatic final compounds (**54**). Quinazoline derivatives with this substitution pattern were not accessible by previous syntheses.

Quaternary carbon formation was utilized in the synthesis of spiro quinazolines (**57**, Scheme 19). The synthetic route started with immobilization of the ethanolamine followed by esterification with an α-amino acid and enabled cleavage of the quinazoline derivatives from the resin without decarboxylation, preserving the quaternary carbon. Synthesis using 2,4-diaminobutyric acid and ornithine introduced the amino group onto the amino acid side-chain and facilitated the formation of spiro [pyrrolidine-3,2′-quinazolin]-2-one (**57a**) and spiro[piperidine-3,2′-quinazolin]-2-one (**57b**) (E. Schutznerova, J. Pospisilova, and V. Krchnak, Unpublished results).

3.3.3 Indoles

Scheme 20 shows the base-triggered sequence of transformations of resin-bound Nos amino ketones (**19**): *C*-arylation (**21**), 5-membered ring closure (**25**), and dehydration to the indazole oxides (**26**). The synthesis of indoles from these common intermediates (**21**) was based on the prevention (or at least minimization) of the N–N bond formation. Reduction of the nitro group in the *C*-aryl derivative

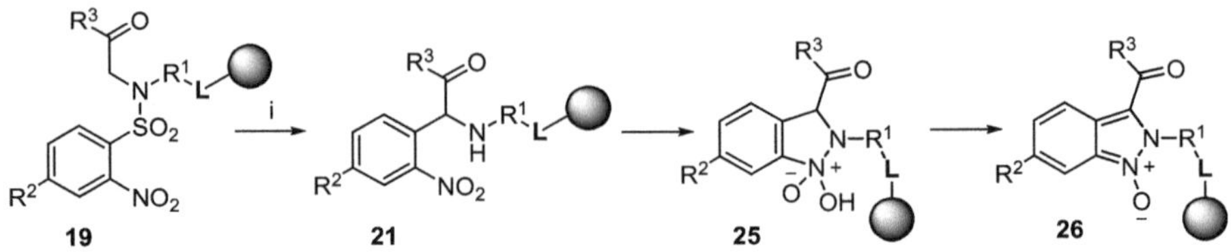

R^1-L = -$(CH_2)_{2\text{-}3}$COO-Wang linker, -$(CH_2)_3$O-Wang linker, -CH(R)COX$(CH_2)_{2\text{-}3}$NH-COO-Wang linker (R = Me, Bn, -CH_2OH; X = O, NH), -CH(Me)CO-piperazinyl-COO-Wang linker
R^2 = -H, -CF_3, -OMe
R^3 = -Ph, 4-OMe-Ph, 4-Cl-Ph

Scheme 20 Base-triggered sequential transformations. Reagents and conditions: (i) 0.1–0.5 M base, DMF or DMSO, rt. to 50°C, 5 min to 48 h

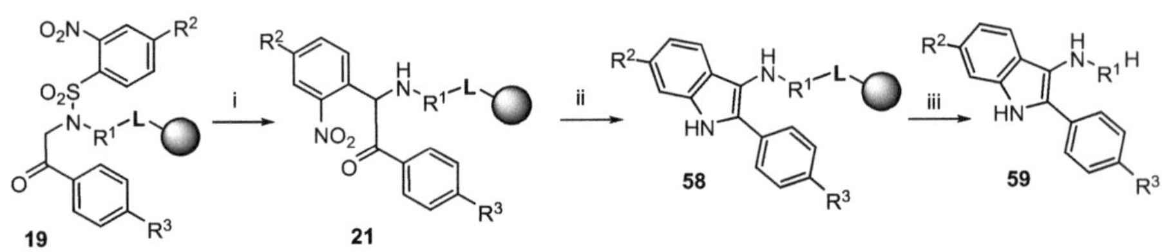

R^1-L = -$(CH_2)_{2\text{-}3}$COO-Wang linker, -$(CH_2)_3$O-Wang linker, -CH(R)COX$(CH_2)_{2\text{-}3}$NH-COO-Wang linker (R = Me, Bn, -CH_2OH; X = O, NH), -CH(Me)CO-piperazinyl-COO-Wang linker
R^2 = -H, -CF_3, -OMe
R^3 = -Ph, 4-OMe-Ph, 4-Cl-Ph

Scheme 21 Synthesis of 3-alkylaminoindoles. Reagents and conditions: (i) base (TEA or DABCO), DMF, rt., 1–16 h; (ii) $Na_2S_2O_4$, K_2CO_3, TBAHS, H_2O/DCM (1:1), rt., 1–16 h; (iii) 50% TFA in DCM, rt., 1 h or TFA/TES/DCM (5:1:4), rt., 1 h

(**21**) and subsequent cyclization was used for the synthesis of another class of heterocycles: the indoles [23]. The critical step was selection of a base and other reaction conditions. The resin-bound intermediate, 2-Nos amide, was exposed to bases of different strength (2,6-lutidine, DABCO, DMAP, TEA, DIEA, DBU, *N,N,N′,N′*-tetramethyl-1,8-naphthalenediamine (a proton sponge), 1,5-diazabicyclo [4.3.0]non-5-ene (DBN), or 2-tertbutylimino-2-diethylamino-1,3-dimethyl-perhydro-1,3,2-diazaphosphorine (BEMP)) to moderate the reaction outcome to favor formation of the *C*-aryl derivative (**21**). In addition to the desired *C*-aryl, unreacted starting material and indazole oxide were also present in the crude product after cleavage from the resin.

The results indicated that weaker bases (lutidine, DMAP) did not trigger *C*-arylation, that TEA and DIEA provided the best conversion to *C*-aryl derivatives (**21**, Scheme 21), that DABCO gave indazole oxides, and that stronger bases (DBU, DBN, BEMP) led exclusively to quinazolines. Therefore, TEA or DABCO was used for the synthesis of indoles (**59**), although the reaction conditions were optimized for each substitution pattern.

3.4 2-Nos Amino Acid Derivatives

N-Alkyl-*N*-Nos amino acid esters and amides represent another class of substrates amenable to base-mediated arylation. Whereas the esters and tertiary amides provided access to *C*-aryl derivatives, the secondary amides yielded *N*-aryl derivatives. Both aryl derivatives represent useful intermediates for the synthesis of heterocycles.

3.4.1 2-Nos Amino Acid Esters and Tertiary Amides

The formation of quaternary carbon derivatives via DBU-mediated C^{α}-arylation was exploited for the synthesis of 3-amino-2-oxindoles (**63**, Scheme 22). The intermediate (**19**) was synthesized on an ethanol amine spacer attached to the Wang resin. This scenario allowed for cleavage of the acyclic precursors (**61**) from the resin (**60**); cyclization to the final heterocycles (**63**) occurred spontaneously after reduction of the nitro group by Zn in solution [24].

3.4.2 2-Nos Amino Acid Secondary Amides

The presence of a proton on the amide nitrogen of the secondary-amide-based substrates (**64**, Scheme 23) was essential for the tandem Smiles-type rearrangement

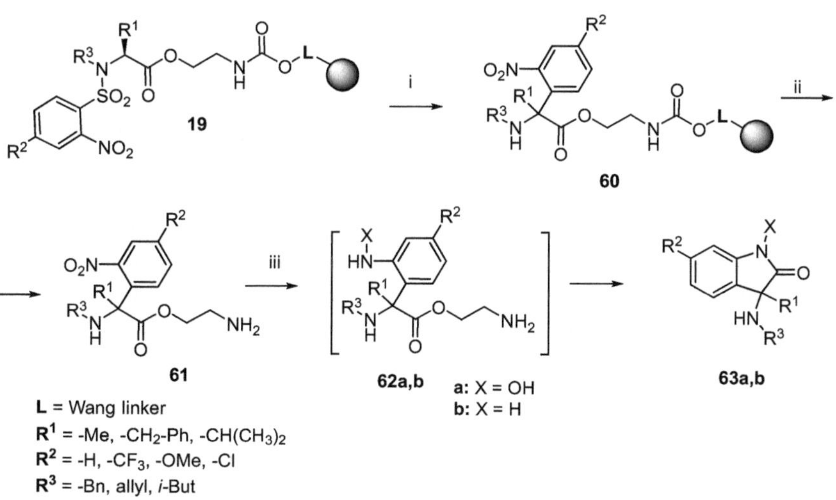

Scheme 22 Synthesis of 3-alkyl-3-(alkylamino)indolin-2-ones. Reagents and conditions: (i) DBU, anhydrous DMF, rt. overnight; (ii) 50% TFA in DCM, rt., 2 h; (iii) Zn/AcOH, rt., 2 h (for X = OH) or 80°C, overnight (for X = H)

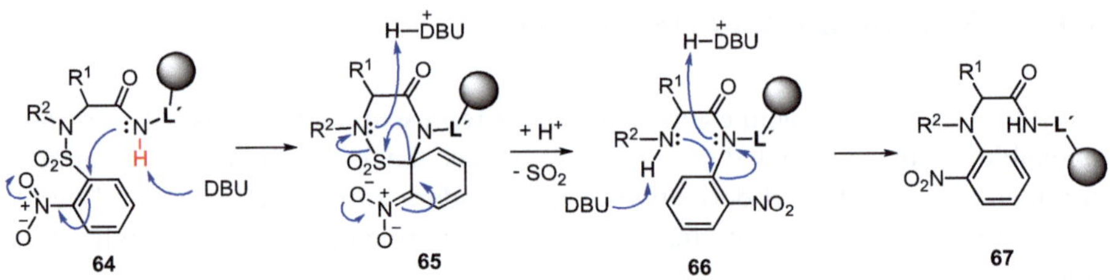

Scheme 23 Mechanism of *N*-arylation of the secondary amide derivatives

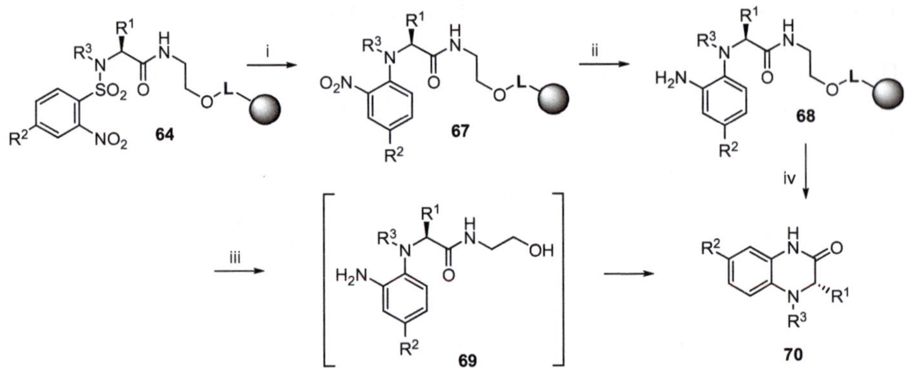

L = Wang linker
$\mathbf{R^1}$ = -Me, $-CH_2-O(tBu)$, $-(CH_2)_4-NH(Boc)$, $-(CH_2)_4COO-tBu$,$-(CH_2)_2COOH$
$\mathbf{R^2}$ = -H, $-CF_3$, -Cl
$\mathbf{R^3}$ = propyl, allyl, propargyl, Boc-aminoethyl, aminoethyl, benzyl, 4-CF_3-Bn, 2,6-diCl-Bn, 2-pyridylmethyl

Scheme 24 Synthesis of 3,4-dihydroquinoxalin-2(1*H*)-ones. Reagents and conditions: (i) DBU, anhydrous DMF, rt., 2–3 days; (ii) $Na_2S_2O_4$, K_2CO_3, tetrabutylammonium hydrogen sulfate, H_2O/DCM (1:1), rt., 1–4 h; (iii) TFA/DCM (1:1), rt., 1 h; (iv) 5% AcOH/DMSO, 70°C, 3 days

and subsequent formation of the *N*-aryl derivatives (**67**), as is shown in the plausible mechanism (Scheme 23).

The base-mediated intramolecular *N*-arylation of secondary-amide-based 2-Nos amides (**64**, Scheme 24) was used for the synthesis of pharmacologically relevant 3,4-dihydroquinoxalin-2(1*H*)-ones (**70**) [25]. The polymer-supported intermediates (**64**) prepared from commercially available Fmoc-protected α-amino acids, 2-Nos chlorides and alcohols underwent base-mediated *N*-arylation. Reduction of the nitro group to an amino group generated the acyclic precursors (**68**), which were subjected to acid-mediated cyclative cleavage.

3.5 2-Nos Benzyl Amines

The last derivatives of Nos amides amenable to arylation are the *N*-benzyl derivatives (**71**, Scheme 25) that can be arylated at the benzylic sp^3 carbon. However, only benzyl amines containing electron-withdrawing groups (Scheme 25, R^3 = NO_2, CF_3, COOMe, 2- and 4-pyridylmethanol) provided *C*-arylation (**72**) at the benzylic sp^3 carbon. By contrast, the arylation was compatible with 2-Nos-derivatives containing both electron withdrawing (CF_3) and electron donating (OMe) groups. Benzhydrylamines (**72**), the products of the intramolecular arylation, were used for the synthesis of indazole oxides (**73**) containing an aryl substituent at the 3-position [26]. An analogous reaction sequence performed using a resin-bound Fmoc-4-aminomethylbenzoic acid provided quinazoline derivatives (**74**).

3.6 Dual Intermediates

To study the relative propensity for *C*-arylation, Smyslova et al. synthesized resin-bound acyclic intermediates containing two potential arylation sites [27]. These model compounds were termed dual intermediates and facilitated evaluation of the direction of the Nos amide-based intramolecular arylation (Scheme 26).

The model compounds had various substitution patterns, types of linker, acidity of the hydrogens on benzyl sp^3 carbon and steric hindrance. The most reactive arylation substrates were 2-Nos amino ketones, followed by 2-Nos amino acid

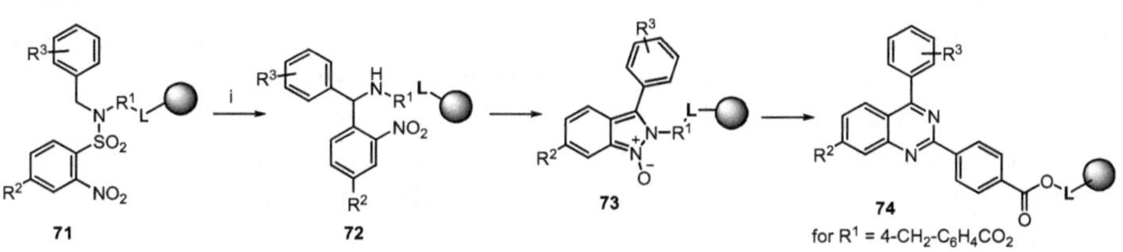

R^1-L = -CH_2O-Wang linker, -CH_2CO-piperazinyl-COO-Wang linker, 4-CH_2-C_6H_4-COO-Wang linker
R^2 = -H, -CF_3, -OMe
R^3 = 4-NO_2, 2-NO_2, 4-CN, 4-CF_3, 4-COOMe, 2-Pyridynyl

Scheme 25 Synthesis of 3-arylindazole oxides and 4-arylquinazolines. Reagents and conditions: (i) DBU, DMF, rt., 1 h to overnight

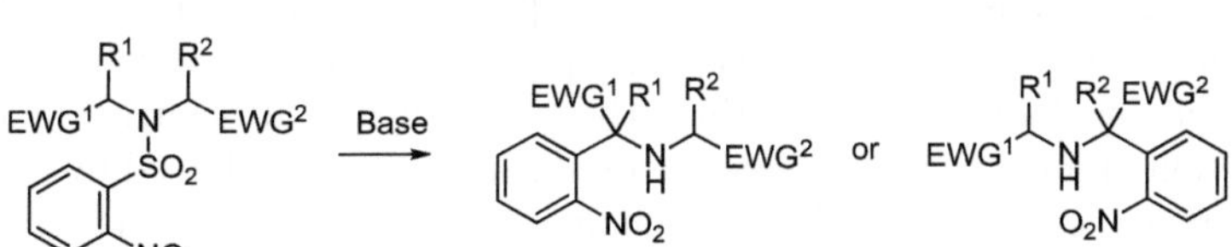

Scheme 26 Dual resin-bound intermediates

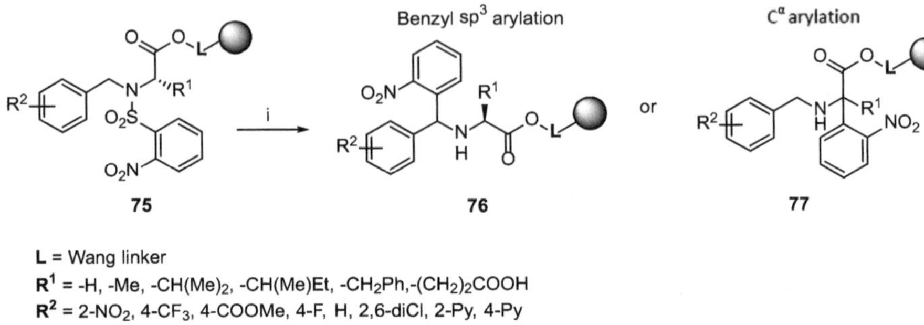

Scheme 27 DBU-mediated arylation of *N*-benzyl amino acid esters. Reagents and conditions: (i) DBU, DMF

esters and tertiary amides. Scheme 27 shows an example of dual intermediates containing *N*-benzyl derivatives of amino acid esters (**75**) representing two positions for potential arylation: the benzylic sp^3 carbon (**76**) and the α-carbon (**77**) of the amino acid. The selection of appropriate substituents on the aromatic ring of the benzyl amine can direct the arylation site. Model compounds prepared using alanine and 2-nitrobenzylamine represented a borderline; a mixture of C^{α} and sp^3 benzylic carbons arylated compounds was obtained. Amino acids with more bulky side-chains yielded sp^3 benzylic carbon arylation products (**76**), whereas alanine derivatives with benzyl amines containing electron withdrawing groups produced C^{α} carbon arylation products (**77**). The least reactive were the *N*-benzyl derivatives containing electron-withdrawing substituents (substrates with electron-neutral and electron-donating substituents did not react). Five tested amino acid esters in combination with a 2-nitro group on benzyl ring provided sp^3 benzyl carbon arylation products (**76**). The alanine ester formed the C^{α} carbon arylation product (**77**) with seven different electron-withdrawing substituents on the benzyl group.

Directing arylation to the benzylic sp^3 carbon yielded benzhydrylamines (**76**), advanced intermediates for the synthesis of quinazolines via indazole oxides. C^{α} carbon arylation (**77**) occurred with alanine as the amino acid and led to the precursors of 3-amino-2-oxindoles.

4 Heterocycles via Nos Nitro Group Reduction

Integration of 2-aminobenzenesulfonamides, prepared by reduction of the Nos nitro group into a heterocycle, has become a frequent route to the nitrogenous heterocycles comprising a fused benzene ring. The typical synthetic scenario includes the construction of an acyclic intermediate on the solid phase followed by nitro-group reduction and cyclization. The reduction was performed with tin(II) chloride dihydrate in early syntheses [28]. Because of the numerous disadvantages of this protocol [29], reduction with dithionate in the presence of a phase transfer catalyst (tetrabutylammonium hydrogen sulfate, TBAHS) was developed [30] and successfully used in recent syntheses. The incorporation of a 2-Nos group into the structure of heterocycles provided fused bicyclic (**78–81**, Fig. 2) [28, 31–33] and tricyclic compounds (**82–86**) (E. Schutznerova, A. G. Oliver, G. Slough, and V. Krchnak, Unpublished results) [29, 34, 35].

4.1 Bicyclic Heterocycles

The synthesis of 2,4-disubstituted 1,2,4-benzothiadiazin-3-one 1,1-dioxides (**78**, Scheme 28) was performed on a BAL linker [28]. The nitro group of the resin-

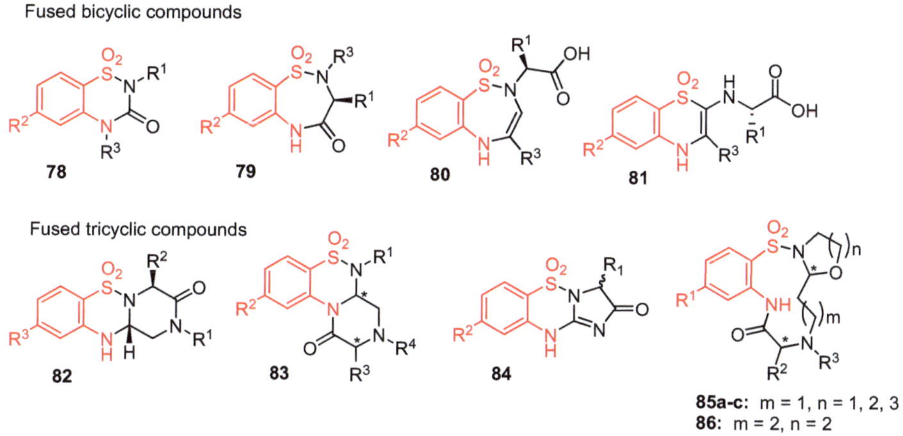

Fig. 2 2-Aminobenzenesulfonamide-derived heterocycles

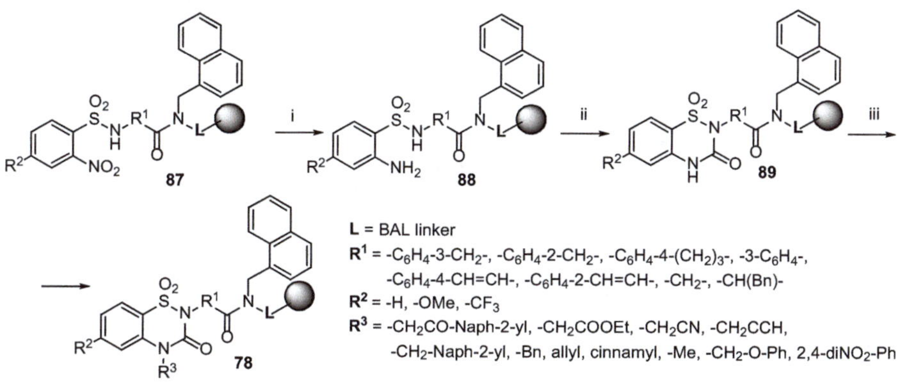

Scheme 28 Synthesis of 2,4-disubstituted 1,2,4-benzothiadiazin-3-one 1,1-dioxides. Reagents and conditions: (i) $SnCl_2{\cdot}2H_2O$, NMP/EtOH, 25°C, 16 h; (ii) CDI, DCM, 25°C, 16 h; (iii) R^3-X, DIEA, NMP, 25°C, 16 h

bound intermediate (**87**) was reduced by tin(II) chloride dihydrate, the six-membered ring (**89**) was closed by CDI and the amide nitrogen was alkylated (**78**).

Seven-membered ring 2,3-dihydrobenzo[*f*][1,2,5]thiadiazepin-4(5*H*)-one 1,1-dioxides (**79**, Scheme 29) were prepared from a resin-bound 2-Nos amino acid alkylated with alcohols (**90**) [33]. After reduction of the nitro group, the acyclic intermediate (**91**) was cleaved from the resin. The 7-membered ring (**79**) was successfully closed by activation of the carboxyl group via acyl chloride. Cyclative cleavage to obtain the product directly from the resin was reportedly unsuccessful.

In an analogous synthesis of other 7-membered heterocycles (**80**, Scheme 30) [31], the final derivatization of the amino group was achieved via the use of bromoketones instead of benzyl alcohols. However, in this case, the methylene

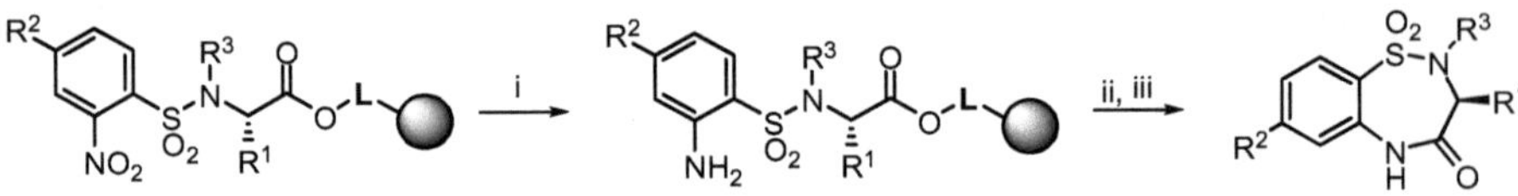

90 **91** **79**

L = Wang linker
R^1 = -Me, -Bn, -CH_2-C_6H_4-4-OH, -$(CH_2)_2COOMe$, -$(CH_2)_2CONHPr$, -$(CH_2)_2$CO-morpholin-4-yl, -CH_2-imidazol-4-yl
R^2 = -H, -OMe, -Cl, -CF_3
R^3 = -Bn, -Et, -$(CH_2)_2N(Et)_2$, -$(CH_2)_3$-Py-4-yl, -$(CH_2)_2$OMe

Scheme 29 Synthesis of 2,3-dihydrobenzo[*f*][1,2,5]thiadiazepin-4(5*H*)-one 1,1-dioxides. Reagents and conditions: (i) $Na_2S_2O_4$, K_2CO_3, TBAHS, DCM, H_2O, rt., 16 h; (ii) 50% TFA in DCM, rt., 1 h; (iii) 20% thionyl chloride, chloroform, 50°C, 1 h

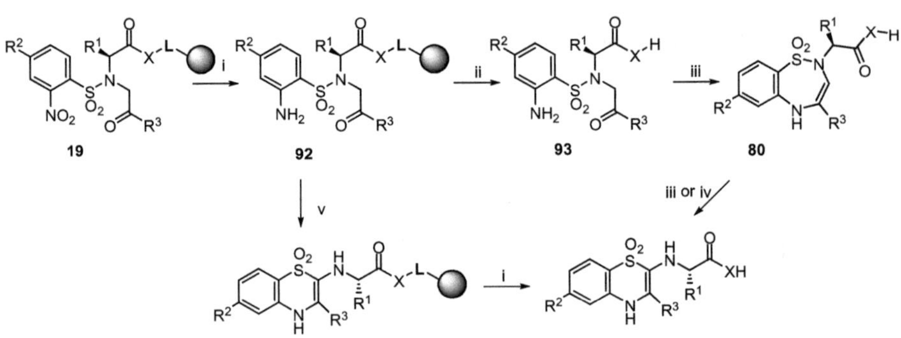

19 **92** **93** **80**

94 **81**

L = Wang linker
X = O, O-$(CH_2)_2$-NH-CO_2
R^1 = -Me, -Bn, -CH_2OH, -CH$(Me)_2$, -$(CH_2)_4NH_2$, -$(CH_2)_2COOH$
R^2 = -H, -OMe, -CF_3
R^3 = -Ph, 4-OMe-Ph, 4-CF_3-Ph, 3,5-diCl-4-NH_2-Ph

Scheme 30 Synthesis of 2,5-dihydrobenzo[*f*][1,2,5]thiadiazepine 1,1-dioxides and benzothiazine 1,1-dioxides. Reagents and conditions: (i) $Na_2S_2O_4$, K_2CO_3, TBAHS, DCM, H_2O, rt., 16 h; (ii) 50% TFA in DCM; (iii) DMSO-d_6, rt.; (iv) DMSO, 70°C, on; (v) 5% AcOH, DMSO, 80°C, 16 h

next to the carbonyl was activated for potential *C*-arylation and indoles and indazoles were also isolated as by-products. The final cyclization was accomplished in DMSO. The target 7-membered compounds (**80**) were stable only in DMSO at lower temperatures. Even at room temperature, slow spontaneous ring contraction to 4*H*-benzo[*b*][1,4]thiazine 1,1-dioxides occurred (**81**) [32], resulting in the formation of a new C–S bond. To facilitate convenient synthesis, this rearrangement was also performed directly on a solid support in 5% acetic acid in DMSO at elevated temperature (**94**).

Model experiments indicated that the ring contraction was triggered by an acid. Acid catalysis was required for the six-membered ring formation, neutral or basic conditions were not effective. The key step of the ring contraction was the attack of sulfur by the alkene electron pair, accelerated by the presence of electron-donating substituents on the aromatic ring (Scheme 31).

4.2 *Tricyclic Heterocycles*

The 2-Nos group was also a key component in the synthesis of fused-ring molecular scaffolds using a tandem cyclic iminium ion cyclization – nucleophilic addition reaction, a very powerful synthetic methodology to access bridged and fused heterocycles [36]. The aromatic amino group, obtained after reduction of the nitro group from 2-Nos amides, served as the nucleophile in both roles: to close the cyclic iminium and also to form the fused ring (Scheme 32).

Stereoselective synthesis of tetrahydrobenzopyrazino-thiadiazinone dioxides (**82**) is shown in Scheme 32 [29]. The intermediates, *N*-acyl-*N*-(2,2-dimethoxyethyl)amines (**97**), were prepared using two different approaches. In route **A**, Rink-resin-supported bromoacetic acid (**95**) was reacted with aminoacetaldehyde dimethyl acetal and the resulting secondary amine (**96**) was acylated with various Fmoc-amino acids (Gly, Ala, Leu, Ile, Ser, Tyr) to afford the key intermediates, *N*-acyl-*N*-(2,2-dimethoxyethyl) amines (**97**). In route **B**, the resin-bound 4-Nos amides underwent Mitsunobu alkylation with glycolaldehyde dimethyl acetal and yielded the alkylated sulfonamides (**99**). Deprotection of the 4-Nos-group followed by acylation with various Fmoc-α-amino acids yielded the *N*-acyl-*N*-(2,2-dimethoxyethyl)amines (**97**). The Fmoc protecting group was replaced by three 2-Nos-Cl's to yield 2-Nos amides (**100**). The nitro group was reduced by dithionite and the resin-bound acyclic intermediates (**101**) exposed to

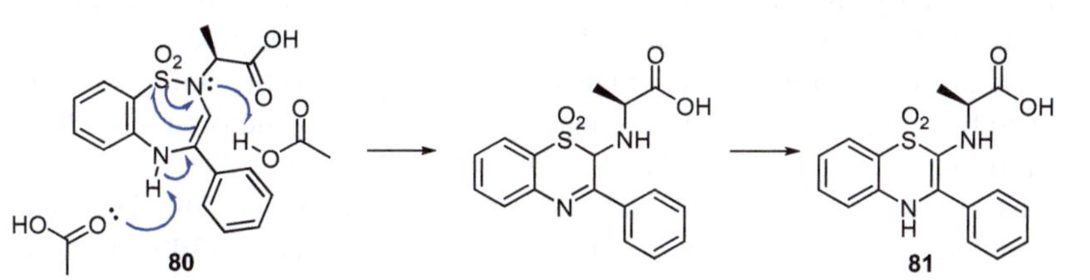

Scheme 31 Mechanism of ring contraction

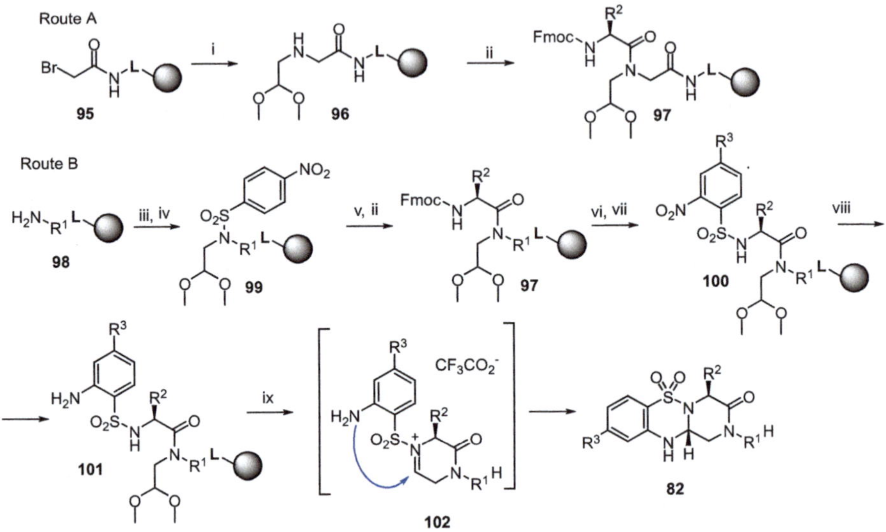

L = Wang, Rink linker

R^1-H = -CH_2COOH, -$(CH_2)_2COOH$, -$(CH_2)_2NH_2$, -$(CH_2)_2CONH(CH_2)_3OH$

R^2 = -H, -Me, -$CH_2CH(Me)_2$, -CH(Me)Et, -CH_2OH, 4-OH-Bn

R^3 = -H, -CF_3, -Cl, -NO_2

Scheme 32 Synthesis of tetrahydrobenzopyrazino-thiadiazinone dioxides via iminium chemistry. Reagents and conditions: (i) aminoacetaldehyde dimethyl acetal, DIEA, DMF, rt., 2 h; (ii) Fmoc-α-amino acid, HOBt, DIC, DCM/DMF (1:1), rt., 16 h; (iii) 4-Nos-Cl, 2,6-lutidine, DCM, rt., 4 h; (iv) glycolaldehyde dimethyl acetal, PPh_3, DIAD, anhydrous THF, 0–50°C, 16 h; (v) 2-mercaptoethanol, DBU, DMF, rt., 5 min; (vi) 50% piperidine in DMF, rt., 15 min; (vii) 4 substituted 2-Nos-Cl, 2,6-lutidine, DCM, rt., 4 h; (viii) $Na_2S_2O_4$, TBAHS, K_2CO_3, DCM/water (1:1), 2 h; (ix) 50% TFA in DCM, rt., 90 min

the cleavage cocktail. The acid-mediated TFA cleavage from the acid-labile linkers with concurrent deprotection of the acetal triggered cyclic iminium formation and subsequent fused-ring closure, yielding the target compounds (**82**).

3,4,4a,5-Tetrahydrobenzo[*e*]pyrazino[2,1-*c*][1,2,4]thiadiazin-1(2*H*)-one 6,6-dioxides (**83**, Scheme 33) were prepared from sulfonamides (**102**) [35]. The nitro group was reduced by dithionite and acylated with α-bromocarboxylic acid under microwave irradiation (**104**). The bromine was replaced by the protected amino aldehyde. Finally, the amino group was derivatized to afford the resin-bound acyclic precursor (**105**). Resin **105** was treated with TFA to cleave the product from the resin and concurrently unmask the aldehyde and trigger cyclic iminium formation followed by nucleophilic addition, yielding the target bicyclic product (**83**).

Synthesis of anagrelide sulfonyl analogues (**84**, Scheme 34) was reported by McMaster and colleagues [34]. The key transformation was the DIC-mediated six-membered ring closure of the Fmoc-thiourea derivative (**108**). Cleavage of the Fmoc group triggered cyclative cleavage of the product (**84**) from the resin.

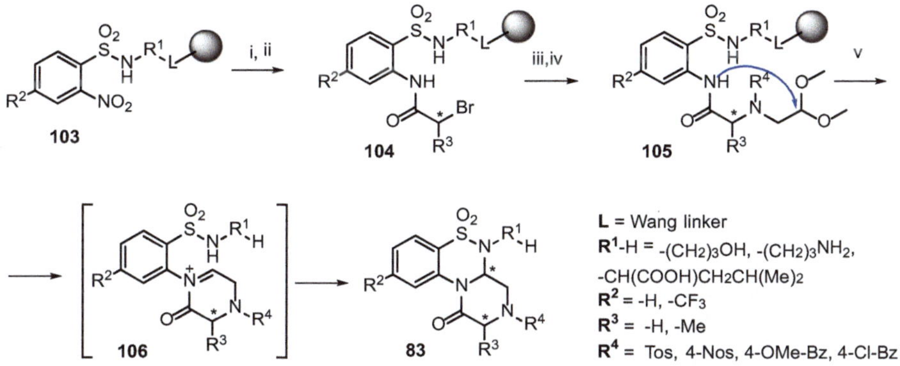

Scheme 33 Synthesis of the fused ring system scaffold. Reagents and conditions: (i) $Na_2S_2O_4$, K_2CO_3, tetrabutylammonium hydrogen sulfate (TBAHS) in water/DCM (1:1), rt., 2 h; (ii) α-bromocarboxylic acid, DIC, THF, 86°C, MW, 1 h; (iii) aminoacetaldehyde dimethyl acetal, DIEA, DMF, rt., 2 h; (iv) Tos-Cl, 2,6-lutidine, DCM, rt., 4 h, or benzoic acid, DIC, HOBt, DMF/DCM (1:1), rt., overnight; (v) 50% TFA in DCM, rt., 90 min or overnight

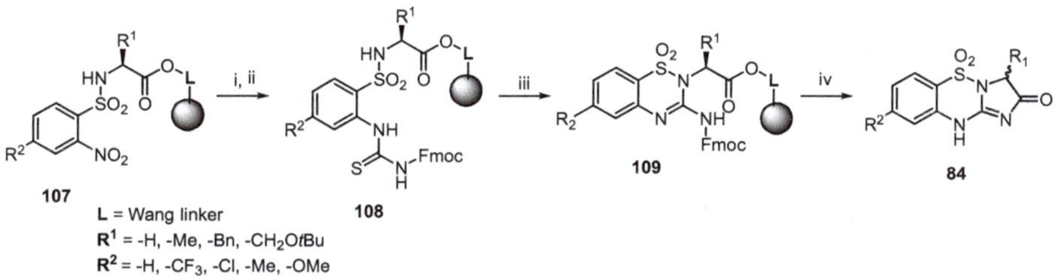

Scheme 34 Synthesis of anagrelide sulfonyl analogues. Reagents and conditions: (i) $Na_2S_2O_4$, K_2CO_3, tetrabutylammonium hydrogen sulfate (TBAHS) in water/DCM (1:1), rt., 2 h; (ii) Fmoc-NCS, THF, overnight; (iii) DIC, DMF, rt., overnight; piperidine, DMF, rt., 30 min

In the synthesis of medium-sized 11- (**85a–c**, Scheme 35) and 12-membered rings (**86**), 2-aminobenzenesulfonamide played crucial role of a conformational constrain, that facilitated population of conformers favorable for the cyclization to larger rings (E. Schutznerova, A. G. Oliver, G. Slough, and V. Krchnak, Unpublished results). Fmoc-amino-alcohols of different chain length (2–4 carbons) were attached to Wang resin and the Fmoc group was unmasked to provide polymer-supported amines (**110**). Reaction with different 2-nitrobenzensulfonyl chlorides (2-Nos-Cl's, R^2) yielded resin **111**. Subsequently, nitro group was reduced by sodium dithionite procedure and obtained aniline derivative (**112**) was acylated with α-bromocarboxylic acids (R^3) under microwave conditions. Acylation was followed by nucleophilic displacement of the bromine with protected amino aldehyde and the last step of building linear precursor **115** included derivatization of secondary amine with aromatic sulfonyl chloride, aryl fluoride or aromatic carboxylic acid. The acyclic resin-bound intermediate (**115**) underwent TFA-mediated unmasking of aldehyde and cleavage from resin. Successful

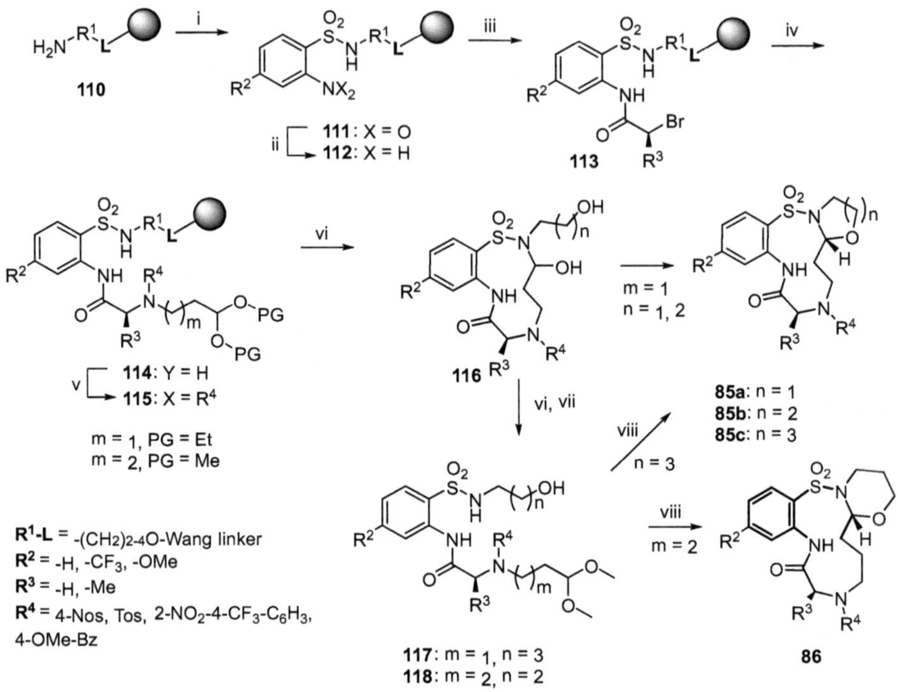

Scheme 35 Synthesis of medium and large rings. Reagents and conditions: (i) 2-Nos-Cl, 2,6-lutidine, DCM, rt., 2 h; (ii) $Na_2S_2O_4$, K_2CO_3, tetrabutylammonium hydrogen sulfate (TBAHS), DCM/H_2O (1:1), rt., 2 h, (iii) α-bromocarboxylic acid, DIC, DCM, MW: 86°C, 50 W, 10 min or 50°C, 50 W, 1 min; (iv) 3,3-diethoxypropan-1-amine or 4,4-dimethoxybutan-1-amine, DIEA, DMF, rt., 2 h (v) 4-Nos-Cl or Tos-Cl, 2,6-lutidine in DCM, rt., on; or benzoic acid, DIC, DCM/DMF (1:1), rt., on; or 4-fluoro-3-nitrobenzotrifluoride, DIEA, DMSO, MW: 80°C, 20 W, 5 min; (vi) 50% TFA in DCM, rt., 1 h; (vii) MeOH, rt., on; (viii) 3.5 equiv. $Sc(OTf)_3$, anhydrous DCM, 0°C, 30 min, then rt., on

cyclization to fused scaffold (**85a–b**) required formation of iminium via hemiaminal (**116**) and subsequent reaction with internal nucleophile – oxygen on the hydroxyalkyl chain. The target fused rings 11 + 5 (**85a**) and 11 + 6 (**85b**) were formed with full stereocontrol of the newly generated chiral carbon. Substrates with longer hydroxyalkyl chain ($n = 4$) or longer protected aldehyde chain ($m = 2$) did not cyclize. In the case of unsuccessful TFA-mediated cyclization the hemiaminals (**116**) were treated with methanol and dimethyl acetals (**117, 118**) were isolated. Scandium triflate was found to be an efficient reagent to catalyze the cyclization, dimethyl acetals (**117**) prepared using 4-aminobutanol gave 11 + 7 fused rings, dimethyl acetals (**118**) with extended chain of protected aldehyde yielded 12 + 6 cycles (**86**).

5 Conclusion

2-Nitrobenzenesulfonamide-containing derivatives represent advanced intermediates in the synthesis of various nitrogenous heterocycles and represent a viable route to solid-phase synthesis of numerous heterocyclic scaffolds regarded as privileged structures. The key transformation is a migration of the aryl group via intramolecular arylation and also reduction of the nitro group and incorporation into a heterocyclic product. Examples of how minor changes in the structure of intermediates and reaction conditions can lead to different outcomes of heterocyclic reactions have been illustrated.

Acknowledgement This research was supported by the Department of Chemistry and Biochemistry, University of Notre Dame, by National program of Sustainability (LO1304), and research grant 16-06446S from the Czech Science Foundation (GACR).

References

1. Fukuyama T, Jow CK, Cheung M (1995) 2- and 4-nitrobenzenesulfonamides: exceptionally versatile means for preparation of secondary amines and protection of amines. Tetrahedron Lett 36(36):6373–6374
2. Biron E, Kessler H (2005) Convenient synthesis of N-methylamino acids compatible with fmoc solid-phase peptide synthesis. J Org Chem 70(13):5183–5189
3. Miller SC, Scanlan TS (1997) Site-selective N-methylation of peptides on solid support. J Am Chem Soc 119(9):2301–2302
4. Lemrova B, Soural M (2012) Solid-Phase Synthesis of 4,7,8-Trisubstituted 1,2,3,4-Tetrahydro-benzo[e][1,4]diazepin-5-ones. ACS Comb Sci 14(12):645–650
5. Fulopova V, Gucky T, Grepl M, Soural M (2012) Solid-phase synthesis of trisubstituted benzo [1,4]-diazepin-5-one derivatives. ACS Comb Sci 14(12):651–656
6. Fulopova V, Soural M (2015) Mining the chemical space: application of 2/4-nitrobenzenesulfonamides in solid-phase synthesis. ACS Comb Sci 17(10):570–591
7. Kleb KG (1968) New rearrangement of the smiles reaction type. Angew Chem Int Ed Engl 7 (4):291
8. Meng Q, Thibblin A (1997) Mechanisms of solvent- and base-promoted imine-forming elimination reactions. J Am Chem Soc 119(6):1224–1229
9. Bowman WR, Coghlan DR (1997) A facile method for the N-alkylation of α-amino esters. Tetrahedron 53(46):15787–15798
10. Wilson MW, Ault-Justus SE, Hodges JC, Rubin JR (1999) A facile rearrangement of N-alkyl, N-(o or p-nitrophenylsulfonamide)-α-amino esters. Tetrahedron 55(6):1647–1656
11. Lupi V, Penso M, Foschi F, Gassa F, Mihali V, Tagliabue A (2009) Highly stereoselective intramolecular α-arylation of self-stabilized non-racemic enolates: synthesis of α-quaternary α-amino acid derivatives. Chem Commun (33):5012–5014
12. Bouillon I, Zajicek J, Pudelova N, Krchnak V (2008) Remarkably efficient synthesis of 2H-indazole 1-oxides and 2H-indazoles via tandem carbon-carbon followed by nitrogen-nitrogen bond formation. J Org Chem 73(22):9027–9032
13. Muth CW, Abraham N, Linfield ML, Wotring RB, Pacofsky EA (1960) Condensation reactions of a nitro group. II. Phenanthridine 5-oxides and benzo[c]cinnoline 1-oxide. J Org Chem 25:736–740

14. Pudelova N, Krchnak V (2009) Efficient traceless solid-phase synthesis of 3,4-dihydropyrazino [1,2-b]indazoles and their 6-oxides. J Comb Chem 11(3):370–374
15. Koci J, Krchnak V (2010) Solid-phase synthesis and chemical properties of 2-(2-amino/ hydroxyethyl)-1-aryl-3,4-dihydropyrazino[1,2-b]indazol-2-iums. J Comb Chem 12(1):168–175
16. Koci J, Oliver AG, Krchnak V (2010) Unprecedented rearrangement of 2-(2-aminoethyl)-1-aryl-3,4-dihydropyrazino[1,2-b]indazole-2-ium 6-oxides to 2,3-dihydro-1H-imidazo[1,2-b] indazoles. J Org Chem 75(2):502–505
17. Schutznerova E, Krchnak V (2016) N-oxide as an intramolecular oxidant in the baeyer-villiger oxidation: synthesis of 2-alkyl-2h-indazol-3-yl benzoates and 2-alkyl-1,2-dihydro-3H-indazol-3-ones. J Org Chem 81(9):3585–3596
18. Krupkova S, Slough GA, Krchnak V (2010) Synthesis of quinazolines from N-(2-nitrophenylsulfonyl)iminodiacetate and alpha-(2-nitrophenylsulfonyl)amino ketones via 2H-indazole 1-oxides. J Org Chem 75(13):4562–4566
19. Boehm HJ, Flohr A, Stahl M (2004) Scaffold hopping. Drug Discov Today Technol 1 (3):217–224
20. Schneider G, Neidhart W, Giller T, Schmid G (1999) "Scaffold-Hopping" by topological pharmacophore search: a contribution to virtual screening. Angew Chem Int Ed 38 (19):2894–2896
21. Bonanomi M, Palazzo G (1977) Products of the reaction between 1H-indazol-3-ol and ethyl chloroacetate. Farmaco Sci 32(7):490–501
22. Fulopova V, Cziesla L, Fleming M, Lu Y, Voelker A, Krchnak V (2015) Traceless solid-phase synthesis of trisubstituted quinazolines. ACS Comb Sci 17(8):470–473
23. Schutznerova E, Krchnak V (2015) Solid-phase synthesis of 2-aryl-3-alkylamino-1H-indoles from 2-nitro-N-(2-oxo-2-arylethyl)benzenesulfonamides via base-mediated C-arylation. ACS Comb Sci 17(2):137–146
24. Gimenez-Navarro V, Volna T, Krchnak V (2015) 3-alkyl-3-(alkylamino)indolin-2-ones via base-mediated C-arylation of 2-nitrobenzenesulfonamides. ACS Comb Sci 17(8):433–436
25. Carbain B, Schuetznerova E, Pribylka A, Krchnak V (2016) Solid-phase synthesis of 3,4-dihydroquinoxalin-2(1H)-ones via the cyclative cleavage of N-arylated carboxamides. Adv Synth Catal 358(5):701–706
26. Kisseljova K, Smyslova P, Krchnak V (2014) Benzhydrylamines via base-mediated intramolecular sp3 C-arylation of N-benzyl-2-nitrobenzenesulfonamides-advanced intermediates for the synthesis of nitrogenous heterocycles. ACS Comb Sci 16(10):573–577
27. Smyslova P, Kisseljova K, Krchnak V (2014) Base-mediated intramolecular C- and N-arylation of N,N-disubstituted 2-nitrobenzenesulfonamides: advanced intermediates for the synthesis of diverse nitrogenous heterocycles. ACS Comb Sci 16(9):500–505
28. Makino S, Nakanishi E, Tsuji T (2003) Efficient synthesis of 2,4-disubstituted 1,2,4-benzothiadiazin-3-one 1,1-dioxides on solid support. J Comb Chem 5(1):73–78
29. Cankarova N, La Venia A, Krchnak V (2014) Polymer-supported stereoselective synthesis of tetrahydrobenzopyrazino-thiadiazinone dioxides via N-sulfonyl iminiums. ACS Comb Sci 16 (6):293–302
30. Kaplanek R, Krchnak V (2013) Fast and effective reduction of nitroarenes by sodium dithionite under PTC conditions: application in solid-phase synthesis. Tetrahedron Lett 54:2600–2603
31. Fulopova V, Krchnak V (2014) Solid-phase synthesis of trisubstituted 2,5-dihydrobenzo[f] [1,2,5]thiadiazepine 1,1-dioxide derivatives. ACS Comb Sci 16(8):412–420
32. Fulopova V, Krchnakova A, Schutznerova E, Zajicek J, Krchnak V (2015) Ring contraction of 2,5-dihydrobenzo[f][1,2,5]thiadiazepine 1,1-dioxides: access to 4H-benzo[b][1,4]thiazine 1,1-dioxides. J Org Chem 80(3):1795–1801
33. Trapani P, Volna T, Soural M (2016) Solid-phase synthesis of 2,3-dihydrobenzo[f][1,2,5] thiadiazepin-4(5H)-one 1,1-dioxides with three diversity positions. ACS Comb Sci 18 (6):349–354

34. McMaster C, Fulopova V, Popa I, Grepl M, Soural M (2014) Solid-phase synthesis of anagrelide sulfonyl analogues. ACS Comb Sci 16(5):221–224
35. Schutznerova E, Verdia P, Krchnak V (2016) Fused ring molecular scaffold with 3D architecture for constrained peptidomimetics: polymer-supported stereoselective synthesis of tetrahydrobenzo[e]pyrazino[2,1-c][1,2,4]thiadiazinone 6,6-dioxide via N-acyl Iminiums. ACS Comb Sci 18(10):655–659
36. Maryanoff BE, Zhang HC, Cohen JH, Turchi IJ, Maryanoff CA (2004) Cyclizations of N-acyliminium ions. Chem Rev 104(3):1431–1628

Index

MIX
Papier aus verantwortungsvollen Quellen
Paper from responsible sources
FSC® C105338

If you have any concerns about our products,
you can contact us on
ProductSafety@springernature.com

In case Publisher is established outside the EU,
the EU authorized representative is:
Springer Nature Customer Service Center GmbH
Europaplatz 3, 69115 Heidelberg, Germany

Printed by Libri Plureos GmbH
in Hamburg, Germany